# Kulturelle Figurationen: Artefakte, Praktiken, Fiktionen

**Reihe herausgegeben von**

Jörn Ahrens, Universität Gießen, Gießen, Germany

Jochen Bonz, Katholische Hochschule NRW, Münster, Germany

Ulrike Vedder, Humboldt-Universität zu Berlin, Berlin, Germany

Annette Vowinckel, Zentrum für Zeithistorische Forschung, Potsdam, Germany

Kultur gilt – neben Kategorien wie Gesellschaft, Politik, Ökonomie – als eine grundlegende Ressource sozialer Semantiken, Praktiken und Lebenswelten. Die Kulturanalyse ist herausgefordert, kulturelle Figurationen als ebenso flüchtige wie hegemoniale, dynamische wie heterogene, globale wie lokale und heterotope Phänomene zu untersuchen. Kulturelle Figurationen sind Produkt menschlichen Zusammenlebens und bilden zugleich die sinnstiftende Folie, vor der Vergesellschaftung und Institutionenbildung stattfinden. In Gestalt von Artefakten, Praktiken und Fiktionen sind sie uneinheitlich, widersprüchlich im Wortsinn und können doch selbst zum sozialen Akteur werden. Die Reihe »Kulturelle Figurationen: Artefakte, Praktiken, Fiktionen« untersucht kulturelle Phänomene in den Bedingungen ihrer Produktion und Genese aus einer interdisziplinären Perspektive und folgt dabei der Verflechtung von Sinnzusammenhängen und Praxisformen. Kulturelle Figurationen werden nicht isoliert betrachtet, sondern in ihren gesellschaftlichen Situierungen, ihren produktionsästhetischen und politischen Implikationen analysiert. Die Reihe publiziert Monographien, Sammelbände, Überblickswerke sowie Übersetzungen internationaler Studien.

Pınar Gümüş Mantu

# Redefining the Political: Youth Experiences of Collective Action in Turkey

Springer VS

Pınar Gümüş Mantu
Justus Liebig University Giessen
Giessen, Germany

Gießener Dissertation im Fachbereich Sozial- und Kulturwissenschaften

ISSN 2567-4242 ISSN 2625-0896 (electronic)
Kulturelle Figurationen: Artefakte, Praktiken, Fiktionen
ISBN 978-3-658-40564-9 ISBN 978-3-658-40565-6 (eBook)
https://doi.org/10.1007/978-3-658-40565-6



This Springer VS imprint is published by the registered company Springer Fachmedien Wiesbaden GmbH, part of Springer Nature.
The registered company address is: Abraham-Lincoln-Str. 46, 65189 Wiesbaden, Germany

# Acknowledgements

This book was written as a doctoral dissertation during my several years of research and academic work in Giessen and Istanbul. First, I would like to thank my supervisor, Prof. Dr. Jörn Ahrens, who shared his support and expertise generously all these years and enabled me to stay the course throughout this research project, in addition to his contribution to the structure and analysis in this work with his critical comments on several versions. I am also grateful to my second supervisor, Prof. Dr. Demet Lüküslü, whose insights helped this work adopt a more nuanced understanding of youth and politics in Turkey. Thanks are due to Prof. Dr. Andreas Langenohl and Prof. Dr. Thomas Brüsemeister for their valuable comments and sitting on my doctoral defense committee.

This research could not have been conducted without the financial support and the inspiring academic environment provided by the GCSC (International Graduate Centre for the Study of Culture) at the Justus Liebig University Giessen. Istanbul Bilgi University Youth Studies Unit has been a place where I was able to further my expertise and broaden my perspective on youth and politics in Turkey as a youth researcher. I would also like to thank to my publishing editor Dr. Cori Antonia Mackrodt at Springer VS, and the editors of the book series *Kulturelle Figurationen: Artefakte, Praktiken, Fiktionen.*

Finally, I wish to thank all the young participants in this research for sharing their stories with me openly and courageously. Without any doubt, this book would not have been possible without their contribution.

Giessen
August 2022

Pınar Gümüş Mantu

# Contents

# Introduction 1

The Gezi protests that began in Istanbul in late May 2013 and spread rapidly across the country were one of the largest protest movements in the contemporary history of Turkey. They were unexpected and extraordinary, especially in terms of the great variety of participants who came from a wide range of political as well as socio-economic and cultural backgrounds. The mass participation of young people, who had often been referred to in public discourse as the "apolitical youth generation", attracted particular public attention. Their ways of protesting via humor and creative performances shaped the distinctive character of the protests, thus making youth participation even more visible. In this context, the protests yet again made it obvious that evaluating young people's politics using old binary oppositions, such as apolitical-political or disengaged-engaged, is unsuited to understanding young people's relation to politics in Turkey.[1] Accordingly, how these crowds of young people, who had long been perceived as uninterested and apathetic, suddenly transformed into active participants in a mass protest movement has emerged as one of the prominent questions related to the topic of the Gezi protests.

This research has been prompted by a desire to understand and critically scrutinize, from a sociological point of view, this "sudden politicization" of the

[1] Young people's lack of participation or interest in traditional politics has been one of the main topics in the field of youth studies for some time now, especially in the subfield of youth political participation. The binary view that labels young people as apolitical and apathetic due to their non-participation in conventional political processes has been criticized against the background of the dynamics and circumstances of young people's lives in European societies in the late modernity (among others, Rys Farthing 2010). Demet Lüküslü (2009) has critically discussed the notion of the apolitical post-1980 generation in Turkey and conceptualized young people's non-participation and conscious withdrawal from politics as 'necessary conformism'. A review of existing research on this topic is provided in Chap. 7.

 
P. Gümüş Mantu, *Redefining the Political: Youth Experiences of Collective Action in Turkey*, Kulturelle Figurationen: Artefakte, Praktiken, Fiktionen,
https://doi.org/10.1007/978-3-658-40565-6_1

young people in the Gezi protests, therefore it focuses on the dynamics of the social context in which being young in Turkey was shaped during the protests, as these contextual aspects relate to the participation of young people. Accordingly, instead of a narrow focus on politicization and youth participation in the protests, this study adopts a broad perspective and looks at what the Gezi protests reveal in terms of young people's lives in Turkey, taking this as the starting point for further analytical inquiry. From this perspective, the analysis of the causes and dynamics of young people's unexpected presence in the Gezi protests crucially informs the discussion of their ways of relating to the political. This study sees the Gezi protests, a momentous social movement, as a case that has the potential to uncover various complex linkages between youth experience, collective action, and politics in Turkey.

The Gezi protests started on May 27, 2013 in Istanbul, as a small group of protesters stood up to the destruction of the trees in Gezi Park in Taksim Square, one of several public green areas in the city center.[2] The urban regeneration agenda for Gezi Park was actually part of a larger urban renewal project titled The Taksim Square Pedestrianization Project,[3] which included the reconstruction of other places in and around Taksim. The first to resist the demolition of Gezi Park were representatives of the Taksim Solidarity Platform,[4] an umbrella association of political organizations and civil society groups that had been raising concerns about the destructive effects of the urban renewal projects in the Taksim area since 2012. As pictures and videos of police interventions started to circulate widely on the social media, many more people, even representatives of political parties, were prompted to join in with the protesters. Solidarity protests started to spread to other cities in the country. In spite of the ongoing police intervention and the rising numbers of injured protesters, on June 1 the movement turned into a mass protest with thousands of people on the streets of Taksim walking toward Gezi Park and were allowed to enter it at the end of the day. While international news channels were broadcasting live from the protests, Turkish mainstream media was ignoring the events unfolding on the streets of Istanbul.[5] From June 1 to June 11, the protesters camped at the park and turned it into a collective living

[2] The course of events is based mainly on the following two sources: (Özkırımlı 2014, pp. 143–147) and (Ayata et al. 2013, pp. 7–18).

[3] Original name in Turkish: Taksim Yayalaştırma Projesi.

[4] Original name in Turkish: Taksim Dayanışma Platformu.

[5] One mainstream channel broadcast a documentary on penguins instead of covering the news from the protests. The "penguin affair" was not only a watershed moment for many people to start questioning the trustworthiness of the mainstream media, but has also regularly featured in humorous materials related to the Gezi protests.

space organized around practices of solidarity, setting up a collective kitchen, a shared library, a volunteer-run infirmary, as well as forums for political discussion and strategy building. Then, on June 11, police interventions started again in the park and around Taksim, and on June 15, the park was closed to the public after the police used excessive force to disperse the protesters. Park forums were subsequently organized in Istanbul and several other cities in the country as a continuation of the activities and protests at Gezi Park.

According to the information released by the Ministry of Interior, by the end of June, 2.5 million people participated in the Gezi protests in 79 cities of Turkey. As many as 4900 were taken into custody, while around 4000 protesters were injured.[6] There were also casualties due to excessive use of pepper spray and exposure to different forms of violence (Ayata et al. 2013, p. 28).

In the summer of 2013, after the Gezi protests, Istanbul in particular, and Turkey in general, were stages of much collective action. Some of these street events stemmed directly from the Gezi protests, e.g. when the funerals of casualties turned into demonstrations. Some other forms of collective action tangibly benefited from the fresh experience of the protests in terms of both size and content. For example, the turnout at the July 2013 LGBT Pride Parade in Taksim was one of the highest ever in the country, and the parade reflected the spirit of Gezi by repeating the demands of the Gezi protests.

While the designation "the Gezi protests" refers to a long and intense period of large-scale collective action across several cities in Turkey, this work must focus on young participants in Istanbul protests. Research on the reflections and experiences of the protests elsewhere in the country would enrich the data and broaden the understanding of this subject. However, this study deliberately limits its scope to Istanbul, mainly for two reasons: first, the occupation of Gezi Park is thought to offer a unique opportunity to discuss political action and being young in conjunction. Therefore, it was decided that young participants who had personally experienced the protests in Istanbul would constitute the main group of research participants in this project. Second, since Gezi Park was the starting point of the protests, and since it was crucially important for this work to zoom in on how young protesters witnessed, experienced, and reflected on the origin of the events, centering the study around young people in Istanbul seemed a reasonable choice.

A survey conducted at Gezi Park on 6–8 July 2013 with 4411 participants showed that the average age of the protesters was 28 and the surveyed group was predominantly made up of university graduates. Moreover, 44.4% of them had

[6] For a related news article, see Şardan (2013).

not taken part in any type of protests, marches or sit-ins before the Gezi protests, and 79% were not affiliated with any political party or civil society organization (KONDA 2014, pp. 8 ff.). Preliminary research thus showed that the participants in the protests were mostly educated young people with no previous experience of collective action or political organization.

The mass movement that started with the sit-in at Gezi Park had various different names that often reflected the political viewpoints of those who coined them. For example, "June Uprising" was used to emphasize the revolutionary potential of the protests from a leftist political perspective, while "Gezi Park Events" remarkably ignored the oppositional character of the movement using a rather imprecise term, which in turn reflected a position against the protests. This work prefers to refer to the sequence of collective actions triggered by the reconstruction of Gezi Park as "the Gezi protests." Using this designation not only acknowledges the social movement dimension, but also signals a more neutral stance compared to terms such as "uprising" or "resistance."

This book presents a wealth of in-depth data on young participants in the Gezi protests interpreted from a cultural-sociological perspective. The analysis is necessarily limited due to time constraints as the research was completed as a doctoral project in 2018, and in a sense already refers to "history." However, beyond this sociological account of a period in the past, this study still makes a relevant contribution to our understanding of young people's lives in Turkey as it comprises an elaborate discussion of the experiences of being young and growing up in Turkey in the last two decades, particularly focusing on crosscutting topics such as socio-economic inequalities, gender relations, social hierarchies, generational conflicts, and socio-cultural pressures and limitations. The author hopes that by bringing youth experiences, perspectives, and action into the focus, this work also contributes, in the broadest sense, to an analysis of contemporary Turkish society. Youth and youth collective action has always featured prominently in the public arena in Turkey. Young people were actively engaging with their social and political context in the second half of the 1960s as masses of them were being politicized as a result of the contact with international youth and leftist movements. Likewise, they were active in the post-1980 period that saw a depoliticization trend in several social movements (religious-ethnic groups, feminist, LGBT and environmental groups, anti-war and anti-globalization protests). However, as will be discussed in detail in Chap. 3, research on young people's collective action and participation is demonstrably limited, and they are still considered an undiscovered group in the history of social movements in Turkey. This work examines young people's collective presence in the post-1980 period in Turkey in great detail, with a view to contextualizing their participation in the

Gezi protests. Relatedly, it indicates that the Gezi protests were not an instant politicization of young people, but rather the articulation of several novel ways of relating to the political which have their roots in previous collective initiatives and mobilizations.

## 1.1 Research Questions, Approach, Aims

In order to understand its subject matter, this study poses several "how" questions. How do young people recall their experiences from the Gezi protests? How do they frame their reasons for participation? How do they reflect on the influences of the protests on themselves and on society? How do young people describe being young in everyday life, especially in the context of the Gezi protests? How do the young participants of the protests define the notion of the political and view their relation to politics and the political sphere? Guided by these questions, employing ethnographic methodology, and inspired by cultural and sociological perspectives, this work aims to understand the relationship between youth and politics.

This research is based on qualitative ethnographic fieldwork that was conducted in two phases in April–November 2014 and January–July 2015 and comprised 40 in-depth interviews with young protest participants. Interviewee profiles were selected so as to reflect the variety of the political backgrounds of the participants, and include feminist, environmental, LGBT and right to the city activists, members of nationalist youth organizations and leftist groups, members of oppositional Muslim groups, young volunteers in civil society organizations as well as young people with no previous political or group affiliation. While maintaining a critical stance, the author has strived to practice reflexivity and develop a continuous awareness of the effect of her role as well as the socio-political context in which the research was conducted. The data was analyzed thematically, and the analysis process itself was designed to have a dynamic relationship with the conceptual discussions throughout the research.

This book studies protest by following cultural and performance theory-informed perspectives in the field of social movements (Jasper 2014; Johnston and Klandermans 1995; Eyerman 2006). Accordingly, special emphasis is placed on the protesters' individual motivations, and self-reflections on their participation. The emotional and moral components of young people's protest experiences and their engagement with the collective identity of the protests are examined using a process-based approach (Melucci 1995). Youth is treated as a sociological concept, by underlining its context-based meaning (Bourdieu 1984 tra. 1993; Durham 2000). It is therefore acknowledged that the definition of youth varies greatly according to where, when, and under which social, economic, and political conditions being young is experienced. This book, and the way it explores the conditions of being young in relation to young people's politics, is also informed by the concept of social generation (Furlong et al. 2011; Wyn and Woodman 2006) that refers to people of similar ages with collective experiences in a particular socio-historical context. Questioning the dichotomies commonly relied on for interpreting young people's relation to politics, such as political vs. apolitical, this work looks for alternative ways of understanding young people and their politics. Accordingly, the concept of the political, without a predetermined definition, is discussed in its broadest sense to allow young people to explain how they understand it. This research therefore proposes to develop a youth-informed definition of the political (O'Toole et al. 2003). Within this framework, the book aims to generate an understanding of the changing and complex conditions of being young, as well as the novel ways in which young people engage in doing politics in Turkey. The historical roots of youth collective action in Turkey, as well as the 2000s wave of global protest movements from the Middle East to the U.S. and Europe, are presented here as the contextual background of the Gezi protests. The book also addresses the socio-economic circumstances and the political atmosphere in Turkey at the time that had a substantial influence on the experience of being young. In addition, young people's experiences and their relations to politics are studied with reference to broader structural processes such as individualization (Beck and Beck-Gernsheim 2002), the dynamics of risk society (Beck 1992), and globalization.

By focusing on the links between the social conditions of being young and collective action, this research seeks to make a contribution at the intersection of two academic fields: sociology of youth and the study of social movements. More specifically, by presenting a qualitative ethnographic investigation of a group of protesters in a historically significant movement, this research adds to the literature on the Gezi protests and the recent wave of global protests in the last decade in general. As it also suggests new perspectives in understanding the relation between young people and politics, this study also potentially deepens the current discussions on youth and politics, specifically in the field of the political participation of young people.

## 1.2 The Structure of the Book

After the introduction, the second chapter extensively discusses the methodology employed in this research. It articulates this work's ethnographic approach, its qualitative methodological premises, as well as the grounds for adopting these particular methodological perspectives. This is followed by a comprehensive description of the fieldwork, including an elaboration of the use of in-depth ethnographic interviews as the main tool of this research, a discussion of the data analysis, as well as how the main themes of the research emerged. Lastly, this chapter contains reflexive notes on the research process and the limitations of the fieldwork. The chapters following Chap. 2 are organized into three parts focusing on protest, youth, and the political.

PART 1, comprising Chap. 3 and 4, aims to provide an exhaustive, theoretically informed and contextualized picture of young people's participation in the Gezi protests. Chap. 3 clarifies why this research views the Gezi protests as a social movement. More precisely, it expounds on the conceptual framework and the literature that form the basis of the empirical discussion of young people's protest experiences. Chap. 4 presents an in-depth analysis of young interviewees' protest experiences.

PART 2, comprising Chap. 5 and 6, scrutinizes the topic of youth with a view to portraying the current circumstances faced by young people in Turkey and the ways they are manifested in the context of the Gezi protests. Chap. 5 sets forth how this research studies the concept of youth, and provides insights on the social conditions of being young both on the local and the global level. It lays the foundation for the empirical discussion in Chap. 6 in which young interviewees' perceptions of what being young means are explored in detail.

PART 3, consisting of Chap. 7 and 8, is centered around the concept of the political and tries to develop an in-depth understanding of young people's politics emerging from the Gezi protests. Chap. 7 reviews critical discussions on youth political participation and presents this work's approach to studying the concept of the political. It constitutes the framework for the analysis of young people's relations with the political, which is given in Chap. 8.

Chap. 9 serves as the conclusion, recapitulating the overall research findings. It presents the general conclusions of the study and describes its contribution to the field. It acknowledges its limitations and includes notes for further research, highlighting the new perspectives and questions that have emerged from this work.

## References

Ayata, G., Çağlı, P., Elveriş, İ., Eryılmaz, S., Gül, İ. I., Karan, U., Muratoğlu, C., Taboğlu, E., Tokuzlu, L. B., & Yeşiladali, B. (2013). Olay Örgüsü. In *Gezi Parkı Olayları: İnsan Hakları Hukuku ve Siyasi Söylem Işığında Bir İnceleme* (pp. 7–18). Istanbul: Bilgi Üniversitesi Yayınları.

Beck, U. (1992). *Risk society: Towards a new modernity* (Mark Ritter, Trans.). Los Angeles, London, New Delhi, Singapore, Washington DC: Sage Publications. (Original work published 1986)

Beck, U., & Beck-Gernsheim, E. (2002). *Individualization: Institutionalized individualism and its social and political consequences.* London, Thousand Oaks, New Delhi: Sage Publications.

Bourdieu, P. (1993). Youth is Just a Word, In *Sociology in question* (R. Nice, Trans.) (pp. 94–101). London, Thousand Oaks, New Delhi: Sage Publications. (Original work published 1984)

Durham, D. (2000). Youth and the social imagination in Africa: Introduction to parts 1 and 2. *Anthropological Quarterly, 73*(3), 113–120.

Eyerman, R. (2006). Performing opposition or, how social movements move. In J. Alexander, B. Giesen, & J. Mast (Eds.), *Social performance: Symbolic action, cultural pragmatics, and ritual* (pp. 193–217). Cambridge: Cambridge University Press.

Farthing, R. (2010). The politics of youthful antipolitics: Representing the 'issue' of youth participation in politics. *Journal of Youth Studies, 13*(2), 181–195.

Furlong, A., Woodman, D., & Wyn, J. (2011). Changing times, changing perspectives: Reconciling 'transition' and 'cultural' perspectives on youth and young adulthood. *Journal of Sociology, 47*, 355–370.

Jasper, J. M. (2014). *Protest: A cultural introduction to social movements.* Cambridge and Malden: Polity Press.

Johnston, H., & Klandermans, B. (1995). The cultural analysis of social movements. In H. Johnston & B. Klandermans (Eds.), *Social movements and culture* (pp. 3–24). Minneapolis: University of Minnesota Press.

KONDA (5. June 2014). *Gezi Report: Public Perception of the "Gezi Protests", Who were the people at Gezi Park?.* Retrieved January 5, 2022, from https://konda.com.tr/wp-content/uploads/2017/03/KONDA_Gezi_Report.pdf.

Lüküslü, D. (2009). *Türkiye'de „Gençlik Miti": 1980 Sonrası Türkiye Gençliği.* Istanbul: İletişim Yayınları.

Melucci, A. (1995). The process of collective identity. In H. Johnston & B. Klandermans (Eds.), *Social movements and culture* (pp. 41–63). Minneapolis: University of Minnesota Press.

O'Toole, T., Lister, M., Marsh, D., Jones, S., & McDonag, A. (2003). Tuning out or left out? Participation and non-participation among young people. *Contemporary Politics, 9*(1), 45–61.

Özkırımlı, U. (2014). Timeline of the protests. In *The Making of a Protest Movement: #occupygezi* (pp. 143–147). London: Palgrave Pivot. (Original source: Amnesty International. (2013). *Gezi Park Protests: Brutal Denial of the Right to Peaceful Assembly in Turkey* (pp. 54–58).)

Şardan, T. (23. June 2013). 2,5 Milyon insan 79 ilde sokağa indi. (2,5 millions of people were on the streets). *Milliyet.* Retrieved November 3, 2020, from http://www.milliyet.com.tr/2-5-milyon-insan-79-ilde-sokaga/gundem/detay/1726600/default.htm.

Wyn, J., & Woodman, D. (2006). Generation, youth, and social change in Australia. *Journal of Youth Studies, 9*(5), 495–514.

# Methodology 2

This work seeks to develop a detailed and in-depth understanding of young people in Turkey by focusing on the case of the Gezi protests. By exploring the topics of protest, politics, and youth, my discussion aims to give a sense of young participants' experiences of the Gezi protests, their relations to the political sphere, and their reflections on being young. My research is grounded in qualitative fieldwork, relying mainly on in-depth ethnographic interviews. The questions of why young people took to the streets and how their protest experiences related to their notions of politics and the political, and to their perceptions of what being young is like, have been the leading concerns of this field research. Being guided by questions of "why" and "how," and directed toward discovering the relations, processes, and meanings attributed by young people to the protests, the political, and youth, this work attempts to produce "deep knowledge" of the young participants of the Gezi protests by examining them in the context of their complex relations (Bray 2008, p. 298).

This chapter focuses on the methodology of the research. Its aim is not only to describe in detail the methods used in the research, but also to discuss the ways in which they are used. Therefore, the methodological approach is not treated as a separate discussion, but rather as an indispensable part of the epistemological and ontological premises of the research. In their discussion on the different approaches in social sciences Donatella della Porta and Michael Keating underline how methodology, which is about the methods used, but not only that, is critically important in terms of the general approach of a study:

> Methods are no more than ways of acquiring data. Questions about methods do, however, come together with epistemology and theory in discussions about *methodology,* which refers to the way in which methods are used. (Della Porta and Keating 2008, p. 28 emphasis in original)



P. Gümüş Mantu, *Redefining the Political: Youth Experiences of Collective Action in Turkey*, Kulturelle Figurationen: Artefakte, Praktiken, Fiktionen,
https://doi.org/10.1007/978-3-658-40565-6_2

This chapter explains the choice of qualitative methodology and ethnographic approach for this study. It outlines the in-depth ethnographic interviews and provides a detailed summary of how interviews with the young people were planned and conducted. This is followed by a description of the data analysis phase and the process of selecting topics and subtopics for the research. The chapter goes on to examine reflexivity as a concept and practice inherent in ethnographic methodology. This is where the positioning of the researcher and the context in which the research was conducted, as well as the young participants' perceptions of the research, come into focus. Lastly, the limitations of the fieldwork are examined in order to paint a coherent picture of the research process, including its drawbacks, and also to propose some points for further methodological discussions.

## 2.1 Qualitative Methodology

Quantitative studies published immediately after the Gezi protests provided some early insights into the profiles of the participants, their motives, and political backgrounds (KONDA 2014; Ercan Bilgiç and Kafkaslı 2013). However, these works, as either online or face-to-face survey-based studies, only offered some introductory findings and did not initiate any debate about the dynamics or the politics of the protests. Moreover, the methodological and scientific credibility of many of these studies has been called into question (Yavuz 2014). Another quantitative study that has a greater explanatory potential critically discusses these early studies and examines their results in comparison with the data collected through an online survey (Erdoğan 2015). Unlike the quantitative studies, recently published qualitative studies based on field research focusing on different aspects of the protests explore the experiences of protestors from different groups, and give clues about the dynamics of the politics they built together (Türkmen 2014; Tanyaş 2015; Gümüş and Yılmaz 2015; Papazian, 2016; Özdüzen, 2017). Qualitative studies that aim to discover the details of the social phenomenon are especially significant in cases such as the Gezi protests, as they were characterized by abruptness, complexity, and pluralism. In addition, qualitative research makes it possible to understand the protests from the perspective of the participants, and therefore lends itself to a more coherent and well-rounded understanding of the protests as social events. Uwe Flick, Ernst von Kardoff, and Ines Steinke underline how qualitative research facilitates collecting data based on the accounts of people reflecting on their own situations:

> Qualitative research claims to describe life-worlds 'from the inside out', from the point of view of people who participate. By so doing, it seeks to contribute to a better understanding of social realities and to draw attention to processes, meaning patterns and structural features. (Flick, von Kardoff, and Steinke 2004, p. 3)

This "inside out" approach, as explained above, is essential to this research, especially with regard to its emphasis on discovering young people's perspectives on their participation in the protests and their relationship with the political. Moreover, this methodological approach highlights the conscious choice made to study a protest movement from the perspective of the protestors, that is to say its main protagonists, not from the perspective of politicians, leaders, or experts.

This work studies young participants in the Gezi protests by looking at their motives for participation, their notions of the political, and their experiences of being young. Without aiming to be representative or to make generalizations about youth in Turkey or the participants in the Gezi protests, this study attempts to discover the details of young protestors' experiences. It seeks to contribute to a better understanding of the protests and young people's politics in Turkey as social realities by centering around young people's lifeworlds. It is also crucial to highlight that the emphasis on the description and exploration of young people's perspectives is coupled with the awareness that qualitative data are representations—not truths—seen through the lens of the researcher and based on his or her field experiences. Underlining the selective processes of viewing and interpretation, Jennifer Mason states that "the elements which a researcher chooses to see as relevant for a description or exploration will be based, implicitly or explicitly, on a way of seeing the social world, and on a particular form of exploratory logic" (Mason 2002, p. 8). Therefore, the descriptions of young people's experiences and lifeworlds are situated in the exploratory logic of the research questions and the conceptual framework of this research. They cannot be evaluated as pure descriptions, but rather as my arguments, primarily based on my close observation and analysis of young people's notions and perceptions of themselves.

### 2.1.1 Ethnographic Approach

Qualitative research, a vast field of inquiry, refers to concepts and assumptions from various scientific traditions such as foundationalism, positivism, post-foundationalism, post-positivism, post-structuralism, and many perspectives and/or methods connected to cultural and interpretive studies (Denzin and Lincoln 2011, p. 2). Among these various research perspectives, it is the ethnographic approach, which is at the heart of qualitative methods in social sciences in relation to the descriptive and interpretive approach (Bray 2008, p. 299), that has guided this study.

By asking "how" questions and looking for extensive answers, qualitative research methods aim to reveal the ways in which the social world is made sense of, experienced, and produced. In doing this, qualitative research zooms in on the existing social reality (Uyan-Semerci 2013, pp. 94–95). Therefore, the social reality, which is usually thought to be obvious, is revisited and reinterpreted through a profoundly qualitative way of observing. In this respect, the most important contribution of qualitative methods to sociological analysis is a set of scientific tools for studying a topic closely and comprehensively. What ethnographic methodology offers is the discovery of the experiences, perceptions, and everyday life practices of individuals and/or groups, and it was this comparative advantage that informed the decision to take an ethnographic, research-based approach to sociological analysis in this research.

Ethnography has a complex history, and its meaning, function, and politics have been continuously redefined in relation to disciplines such as anthropology, sociology, and cultural studies. Its complicated history may be seen as one of the reasons why ethnography does not have a standard, well-defined meaning (Hammersley and Atkinson 2007, p. 2). However, ethnography as an approach implies certain characteristics, such as being naturalistic and holistic, as well as theoretically open and self-reflexive (Bray 2008, pp. 300–303). Its methodology is employed here as not just as a set of tools or ways used to gather data during fieldwork, but as an approach to the research process as a whole. This methodological choice thus reflects the general aims and priorities of the work as well as how epistemological and ontological aspects of the research have been linked with the methodology.

As mentioned above, this study aims to develop an in-depth understanding of how young people relate to the political, informed by their own conceptualizations and reflections. It is intended to take a naturalistic approach toward working with young people, interpreting how they give meaning to their experiences without constructing any abstract setting or predefined categories (Bray 2008, p. 300). Throughout the fieldwork, positioning myself as close as possible to my subjects' way of thinking and perceptions of their own lives and environment, that is to say, putting myself in their shoes, proved to be an important asset in developing a substantial account of their perspectives. The ethnographic approach opens up a possibility of scientifically motivated empathetic attitude: "While scientifically motivated, ethnographic research is carried out with a humanist emphasis, delving empathetically into the complexity of the culture and political world of the people" (Bray 2008, p. 301). From this standpoint, this study stresses the individual accounts and micro-practices of young people. Most importantly, however, the detailed pieces of information based on individual stories are contextualized in a holistic interpretation, which enables a thorough and comprehensive understanding of young people's experiences of protest, politics, and youth based on the particulars of the field data.

The ethnographic approach is open-ended in terms of data collection and the theoretical framework of the research. Approaching his or her subjects with an open mind, the researcher does not search for specific manifestations of already-formed theoretical ideas (Bray 2008, p. 303). Instead, she or he remains open to new questions that may emerge from the field research and continually evaluates how these influence and revise theoretical considerations of the work. The present study, too, employs an intentionally open-ended approach to its theoretical concepts. "The political" is used as a particularly open term; it is defined and described by the young people themselves. This study therefore aims to contribute to the theoretical discourse on new ways of defining the political, especially from a youth-informed viewpoint.[1] "Youth" and "protest" are also taken as experiences and processes that are continuously redefined and rediscovered by and with the young participants of the protests throughout the course of the research.

Self-reflexivity is an inherent part of ethnographic methodology and refers generally to the researcher's systematic practice of continuously reflecting upon his or her position and its effects on his or her relationship with the research participants, as well as on the results of the research itself. The concept of reflexivity also stresses the importance of taking into account the context in which the research takes place; this includes both the researcher's background and the social-political environment of the research:

> What this represents is a rejection of the idea that social research is, or can be, carried out in some autonomous realm that is insulated from the wider society and from the biography of the researcher, in such a way that its findings can be unaffected by social processes and personal characteristics. Also it is emphasized that the production of knowledge by researchers has consequences. At the very least, the publication of research findings can shape the climate in which political and practical decisions are made, and it may even directly stimulate particular sorts of action. (Hammersley and Atkinson 2007, p. 15)

[1] See Chap. 7: Studying Youth and Politics for a detailed discussion on how theoretical openness in terms of "the political" relates to the literature on youth political participation.

With this in mind, reflexivity in this research serves to develop an awareness of both the role of the researcher and the context in which the research is conducted. A detailed account of my reflections on several aspects of the research process is presented later in this chapter.

As an ethnographic study of the young participants of the Gezi protests with a special focus on their relationships with the political, this work relates to different fields such as political ethnography, ethnography of social movements/protest, as well as ethnography of youth culture. This study does not fit neatly into any one of these areas of inquiry, but rather stands at their crossroads. Even if this work does not investigate a closed youth culture community, young people's cultural practices are at its center. In keeping with this, the debates sparked by ethnographic studies of youth cultures have inspired my reflexive practice. Since this study concentrates on how young people experience politics and the political in their everyday lives and practices, it makes a clear reference to the "lived experience of the political," which is defined as one of the ways of studying politics with an ethnographic gaze, by looking at what is "invisible from the non-ethnographic vantage points" but "of consequence to politics in some way" (Baiocchi and Connor 2008, p. 140.) With a rethinking of the existing conceptualizations of the political informed by young people's descriptions and perceptions as its goal, this research also highlights the potential of ethnographic studies to call into question assumptions of traditional political studies, therefore offering a possibility of retheorization (Ibid, p. 141). From the perspective of social movement research methodology, this work investigates how activists view their participation and how they understand their social world; it is therefore an attempt to understand a social movement through a specific group of its participants (Blee and Taylor 2002, p. 95).

## 2.2 Fieldwork

### 2.2.1 Preliminary Field Research

I started my field research in September 2013, just after the Gezi protests in Istanbul in May and June 2013, with four preliminary in-depth interviews with young protest participants. The aim of this preliminary research was to determine in what ways these protests, as a meeting point for many young people from many different political and socio-economic backgrounds, could provide new possibilities for understanding youth, their cultural practices and relationship with the political in Turkey. This early contact with young people focused on how they characterized their participation in the protests. These initial interviews were loosely structured, consisting mainly of open-ended questions, which allowed the interviewees to talk about what was important for them regarding the protests. Within this relatively free structure, interviews were moderated so as to ensure that a summary of the interviewee's biography, their story of protest participation, as well as their reflections on being young in Turkey were included. At this stage, along with in-depth interviews, ethnographic field observations were recorded. Since the protests had concluded not long ago, the post-protest period was observed. Following the end of the occupation period in the park by police intervention, park forums, local initiatives, and neighborhood forums started to emerge. Although the protests centered around Gezi Park had already ended, smaller marches and protests were still taking place occasionally. Being in the field at that time provided me with the opportunity to witness the developments. In addition to the interviews, in spontaneous chats, I had the chance to collect data on the participants' reflections on the protest period as well as on how their participation had changed their everyday lives and their relationship with the public sphere.

The preliminary field research determined the next steps of the research design in two ways: First, it proved that focusing on young people's experiences in the Gezi protests would allow me to discover their cultural practices and lifeworlds, as well as to discuss politics and the political with them. Their participation in the protests seems to have been a life-changing experience for the participants; they referred to the transformative potential of the protests that made them rethink, revise, and reshape their understanding of the political. It was therefore obvious that the way in which their relation to politics changed in this process warranted attention. In addition, their initial accounts also revealed that the protests carried the potential to uncover the diverse meanings young people attributed to the experience of being young. In light of these points, my preliminary research provided a rough idea of the thematic categories that emerged subsequently. Accordingly, the main concepts of the research were defined as follows: protest, the political, and youth. This initial fieldwork guided the reworking of the theoretical framework and the preparation for the next stages of field research.

### 2.2.2 In-Depth Ethnographic Interviews

The fieldwork was completed in two stages, from April to November 2014 and from January to August 2015. The main method was the in-depth ethnographic interview, the primary aim of which was to produce deep and detailed knowledge of young people's experiences. Grounded in the evaluation and rereading of the interviews conducted as part of the preliminary fieldwork, in-depth interviews were planned to cover the focal points of the research; however, no exact questions were formulated that were posed to all interviewees. Instead, a more flexible approach was adopted to allow the discussions to proceed in a more natural way (Hammersley and Atkinson 2007, p. 116). As the fieldwork put a clear emphasis on the voices and concerns of young people, open-ended interviews with minimal intervention enabled the interviewees to say whatever they felt was relevant and important (Bray 2008, p. 310). However, given this flexibility, it was important to keep the discussions focused on the main research topics. I attempted to preserve a balance between open-ended conversations and research-focused discussions throughout the interviews. This balance stresses the fact that ethnographic interviews are not simply "free chats", but structured, flexible, and creative ways of gathering data. Even if they appear to resemble one another, ethnographic interviews "are never *simply* conversations, because the ethnographer has a research agenda and must retain some control over the proceedings" (Hammersley and Atkinson 2007, p. 117, emphasis in original).

In general, most of the interviews started with introductions, in which the interviewee and I got to know each other and established some details about the interviewee's background before discussing his or her participation in the protests. While talking about the protests, the conversation would veer to politics, the political, and the interviewees' experiences of being young. The Gezi protests thus served as the point of departure and main framework within which young people could reflect on being young and political. Although the fieldwork did not include comprehensive participant observation during the protest period (since the interviews were conducted after the protests), the ethnographic gaze was employed, and field data procured through a combination of interviews and ethnographic fieldwork diaries. In keeping with the ethnographic gaze, I did not consider the interviews and participant observation to be separate practices. Rather, there was an element of participant observation in all of the interviews. Framing participant observation as not only a scientific research method, but also as the way we naturally exist in social relations, Martyn Hammersley and Paul Atkinson define the interview as a participant observation activity for both the researcher and the participants:

> Whatever their form, interviews must be viewed as social events in which the interviewer (and for that matter the interviewee) is a participant observer. The ethnographer may be able to play a more dominant role than is the case in other contexts, and this can be capitalized upon, both in terms of when and where the interview takes

place and who is present, as well as through the kind of questions asked. (Hammersley and Atkinson 2007, p. 120)

Thus, given that the interview is a social relationship and the interviewer one of its participants, the role of the interviewer (the researcher) is not stable, but rather changes with the context in which the interview takes place. Even if an interview is always an artificial event, which is specifically organized in line with certain research aims, it is unlike a laboratory setting in which the conditions can be fully controlled; it is, rather, a part of real life with its uncertainties. The value of the method also lies in the fact that observing and analyzing people in detail in their real lives provides unique kinds of data that the researcher could not gather in other ways. Moreover, the interview is a collective activity, shaped by both the interviewer and the interviewee, not one organized and directed solely by the researcher. With such an understanding of the ethnographic interview as a method, this study bears in mind the limited role of the researcher in the interview setting, and casts the interviewer as a participant and observer, not the leader and director of the conversation.

In total (including the preliminary stage), 40 in-depth interviews were conducted with young people who participated in the Gezi protests. Participation is defined as having taken part in the demonstrations and/or the occupation of the park. Some interviewees were active members of local initiatives founded after the protests in addition to having demonstrated and camped in Gezi Park.[2]

Interviewee profiles were primarily created to reflect the variety of political backgrounds of the participants of the protests. Environmental, right to the city, feminist, and LGBT activists were put on the top of the list in light of the substantial visibility of young people from these groups in the protests. In addition, young members from nationalist organizations as well as young people who articulated their political identity mostly with reference to the historical leftist tradition were included in the field research. Furthermore, oppositional Muslim participants were interviewed to get an impression of their experiences of conflict and cooperation with other participants given the predominantly secular atmosphere of the protest environment.[3] Also included was an activist from an anti-militarist and

[2] Exceptions to this definition of participation were two interviewees who had supported the protests in online spaces as they unfolded but became active members of the local initiatives afterward. They were included because the local initiatives are considered here as an integral part of the Gezi Protests.

[3] Here, the phrase "predominantly secular" is used mainly referring to the oppositional character of the Gezi protests with regards to the Islamization process going on in Turkey, which is discussed in Chap. 4.

anti-capitalist initiative, and an independent activist, as their experiences helped represent the variety of political identities involved in the protests. It is important to note that it would be neither possible nor accurate to treat these categories as strictly separate from each other. Rather, they occasionally overlap, as in the case of an oppositional Muslim activist who was a feminist activist at the same time, or a feminist activist also identifying herself as a leftist. Apart from these, participants who did not have any political affiliation but had worked as volunteers in civil society organizations conducting activities with youth groups were included in the interviewee group in order to explore the experiences of young people who were already accustomed to discussions on youth and could therefore potentially assume a critical perspective on being young in Turkey. In addition, young people who did not have any political or civil society affiliation made up a significant part of the participant group as examples of individual participation with no links to any organizations or groups. The group consisted of six environmental, four right to the city, four LGBT, and four feminist activists, three members of nationalist organizations, five leftists, three oppositional Muslims, one member of an anti-militarist and anti-capitalist initiative, one independent activist, three volunteers in civil society organizations working on youth, and six young people with no affiliation. In addition, seven out of 40 were actively involved in the local/neighborhood initiatives formed as a continuation of the protests. One interviewee, in addition to his political affiliation, was also a member of a football fan group which had been active in the protests.

It is important to note here that the research sample did not primarily aim to reflect political affiliations based on ethnic/religious attributions, i.e. youth in Kurdish movements. This was mainly due to the impossibility of analyzing young people's politics in relation to ethnic/religious identities in Turkey without situating those identities in a detailed historical discussion of the struggles and political movements that shaped them, which was obviously beyond the scope of this research.[4] However, in cases where young people declared their ethnic/religious affiliations as a crucial part of their political identities, such as being Kurdish, Armenian, or Alevi, then this was mentioned while describing the interviewee's background and taken into consideration in the analysis. Therefore, Kurdish, Armenian, and Alevi youth are not absent from this research, but included in the sample as cases of intersectional political identities, such as being a LGBT and Kurdish rights activist at the same time.

[4] The politics of Kurdish, Alevi and Armenian young people will be touched upon in the discussion on youth in social movements in Turkey in Chap. 3: Youth and Social Movements.

The age range included in the fieldwork is between 18 and 35, among these, 27 of the interviewees were between 18 and 25, ten were between 26 and 30 and three of them were between 31and 35.[5] Age distribution was designed in accordance with the quantitative data that shows 47.3% of the participants in Gezi Park were 18–25 years old, and 20.3% were 26–30 years old (KONDA 2014, p. 8). Moreover, since the age 18–25 generally corresponds to the period young people spend at university, it was reasonable to include much more young people from this age group given the high level of student participation—one out of every three Gezi Park protesters was a student (KONDA 2014, p. 11). In addition, considering the phenomenon of prolongation of the youth period under the present global socio-economic circumstances, participants aged 26–35 were included in the sample in order to gain a better grasp of the various experiences of being on the verge of adulthood in Turkey.

Twenty-three of the participants were university students (three of them graduates) at various departments of different universities in Istanbul. Three of these students were also employed part time or fulltime. Fourteen of the participants were employed in different fields such as corporate work, the service sector, freelance translation or graphic design, jobs on TV and cinema projects. Three out of the forty young people were neither students nor employed, two of these were looking for jobs while one was preparing for the university entrance exams for the second time. Apart from their occupational status, a specific categorization focusing on the socio-economic status of young people and their families was not made, since the research was not designed to be an analysis based on socio-economic conditions. However, since the interviewees' quotations were analyzed with a holistic view with special attention paid to the integrity of each individual interview and the interconnection of the information it contained, information about socio-economic background of the participants was taken into consideration where it was relevant and applicable.

Representing both sexes equally was not considered a priority when choosing the participants during the fieldwork. Since gender and gender distribution do not occupy a central place in this research, aiming for equal representation of both sexes was deemed unnecessary. Rather, an effort was made to keep a reasonable balance in order to not ignore or overrate experiences of either sex. Accordingly, twenty-three interviewees were female, sixteen were male, and one was transgender.

[5] The initial age range defined for the participants was 18–30. However, in the course of the fieldwork, three interviewees over 30 were included since their experiences were unique and important due to their political background.

I had access to my interviewees mainly thanks to my existing contacts from social movements, volunteer organizations, and university groups. I used the snowball technique to reach out to new interviewees following my draft list of participant profiles. Usually, one interview opened up possibilities for several others, and I chose to pursue the most suitable prospective study participants making a pre-assessment of how the characteristics of the potential interviewee (political affiliation, age, gender, and occupation) matched with the priorities of the fieldwork design.

At the beginning of each interview, the interviewees were orally informed about the ethical considerations and rules of the research, and their informed consent was audio recorded. I ensured that the participants understood the ethical concerns of the research and how their identities would be protected in the research process and in any publications or presentations emerging from this fieldwork. The data collected in the interviews were anonymized as soon as the recordings were transcribed, and participants are cited in the texts only by their political background, age, and occupation or student status. The recordings and transcripts were kept safe and were not shared with third parties for any reason. In this way, I made sure to protect the interviewees from any repercussions.

## 2.3 Data Analysis

The data consisted of full interview transcripts and field notes. I completed the transcripts myself and did all of the analysis work by hand. Since ethnographic methodology requires repeated close reading of all the field materials and a progressive construction of the analysis while working through the data and refining the theoretical framework of the research, I saw analyzing the data manually as crucially important for keeping my contact with the field data close, open, and reflexive.

Extending the fieldwork over time facilitated a dynamic interaction between the data and the theoretical framework of the research. After each stage, while working through the collected data, I simultaneously revisited and revised the theoretical framework. In this way, my investigation aimed at achieving continuous interplay between data and ideas, inspired by Hammersley and Atkinson's discussion of grounded theorizing:

> It was also a rejection of the contrasting conception of research as dredging through an inert mound of data to produce descriptions of what is there. The central injunction of grounded theorizing is that there should be constant interplay between data and ideas

> throughout the research process. Ideas are emergent from one's experience in the field, and from one's preliminary analytic reflections on the data. As this should make clear, emergence is a function of the analytic work one puts in: it does not 'just happen'. (Hammersley and Atkinson 2007, p. 159)

It follows that an analysis is the final product of the researcher's constant work on his or her field experiences whereby they are examined against the researcher's existing ideas and assumptions, as well as the theoretical framework of the research. Analysis is an ongoing process of rethinking and revision. In line with this approach, my data analysis was already in progress from the beginning of the fieldwork and I did not have to wait for all of the data to be collected. As the fieldwork proceeded, the focus of the research progressively narrowed (Hammersley and Atkinson 2007, p. 160).

The data was processed by means of thematic analysis. Richard E. Boyatzis' definition of thematic analysis as "a way of seeing" (Boyatzis 1998, p. 1) informed this research, especially the analysis of the field material. Emphasizing how the researcher "sees" his or her field, this approach points to the researcher, without overlooking his or her subjectivity, as an individual who looks, sees, understands, and analyzes what she or he collects from the field. Therefore, the researcher is not just an expert who applies procedures to produce an analysis based on the field data. Rather, the analysis is the result of the researcher's methodologically guided observation and his or her research questions that are also created by "seeing." In that sense, thematic analysis was quite compatible with the methodology adopted for this field research underlining the awareness of the position of the researcher.

The data analysis primarily aimed to develop themes/categories/concepts through which young people's perceptions and experiences could be fully described and explained. These themes were also examined so as to open up the space for insightful discussion at the intersection of field material and theoretical reflections. Apart from that, the setting of themes was not imagined as a process solely performed by the researcher, but as one in which interviewees, too, participated through their reflexive accounts given in the interviews. Methodologically, while developing the analysis, themes were considered as units that could be used to analytically organize what is seen and more specifically how something is seen:

> If sensing a pattern or "occurrence" can be called *seeing*, then the encoding of it can be called *seeing as*. That is, you first make the observation that something important or notable is occurring, then you classify or describe it. (Boyatzis 1998, p. 4, emphasis original)

Within this framework, protest, youth and the political emerged as the three main themes of the study, to be used as a lens through which to examine young people's reasons for taking to the streets as well as the links between their participation in the protests and their ways of relating to politics and being young. In other words, these themes, as patterns sensed through the field materials, aim to interpret aspects of the phenomenon (Boyatzis 1998, p. 4) of youth in the Gezi protests.

Although the themes crystallized during the fieldwork, it should be noted that they were not merely the products of the field research. Rather, they started to take shape in the earliest stages of the research project. In addition to the data collected, my previous observations, experiences, and previous academic study of the topic decisively influenced the choice of the central themes of analysis. As Grey W. Ryan and Russell H. Bernard put it: "Literature reviews are rich sources for themes, as are investigators' own experiences with subject matter. More often than not, however, researchers induce themes from the text itself." (Ryan and Bernard 2003, 205).

In the first stage, the data was organized around the three main themes of investigation, namely protest, youth, and the political. After that, subthemes for each main theme emerged through repeated and close reading of the material. Subthemes were usually formed around common experiences/phenomena emerging under the umbrella of the main theme. For example, "the Gezi spirit" emerged as a subtheme of protest since it was a phenomenon frequently referred to by young people. However, this subtheme encompassed not only what young people had in common when it comes to their relation to the Gezi spirit, but also the diverse ways in which they related to it, and their contradictory reflections were included in the discussion. The interviewees' accounts were analyzed against the backdrop of their political or group affiliations, or non-affiliations as the case may be, as well as their age, gender, occupation, and socio-economic and cultural circumstances, as long as these were relevant. The following section elaborates on how each main theme was studied through subthemes guided by research questions.

### 2.3.1 Main Themes: Protest, Youth and the Political

The question of how young people experienced the protests guided this research from the beginning. Departing from this question, this work focuses on young people's perceptions of the protests in order to understand the nature of their participation in them. This point of departure enables a closer look at individuals and their ways of engaging with the protests. Young individuals' descriptions of their protest experiences are examined here from a cultural perspective, taking into consideration the participants' thoughts, emotions, and morals. This allowed for more attention to be devoted to emotions and morals, factors that are usually underestimated in analyses of individuals' engagement in collective action. The performative and process-oriented approach makes it possible to decipher how individuals gradually and dynamically formed a collective identity during the protests.

Throughout the interviews young people mostly shared long and detailed narratives of their protest experiences. Usually, asking a simple question such as, "How did you decide to participate in the protests?" resulted in exhaustive accounts covering not only the day they made up their minds to get out on the streets, but also the run-up to the protests, occupation days at the park, and the period after the protests. These descriptions provided background information on the protests, young people's reasons and forms of participation, the collective spirit of the Gezi Park, as well the after-effects of the protests. The accounts of the run-up to the protests are analyzed by stressing the participants' feelings of unhappiness and oppression related to the implications of global neoliberal rule for life in the cities in particular and the government's authoritarian attitude toward youth in general. The violent treatment of the initial protestors in the park was emphasized as a decisive factor that prompted many young people to participate in the protests, and defined as a moment of moral shock. The protesters' collective identity, commonly referred to as the Gezi Spirit, is studied here via the concept of solidarity with emphasis on the feelings of trust, security, and happiness the protesters associate with it. In a similar vein, the experience of camping in the park is highlighted as an opportunity for young people to practice their ideal understandings of social life. Lastly, young people's reflections on what changed after the protests are processed via the concepts of transformation, political consciousness, criticism, disappointment, and trauma.

This field research included an empirical inquiry into the concept of youth, triggered by the question, "Who do we talk about when we talk about 'youth' at the Gezi protests?" Conceptualizing youth as a power relationship from a sociological perspective, the question of how age and, therefore, being young shaped young people's everyday lives was one of the main points investigated in all of the interviews. What these young people had in common is analyzed through the concept of generation as a socio-historical positionality that enables people of similar age to feel and act as a collective. Accordingly, the meanings of being young were studied as young people reflected on their experiences under the socio-economic and political conditions at the time. Situated at the intersection of transition approach and cultural approach to studying youth, the analysis of this theme consistently attempts to trace how young people experience their transition from youth to adulthood under given structural circumstances in several fields of life such as family, education, work, etc., while at the same time an effort is made to look at their responses through a cultural lens.

What fascinated me at first was that most of the interviewees had already developed critical perspectives concerning the function of the notion of youth in their lives. Many of the young participants frequently questioned the common sense universal attributions according to which being young is associated with being energetic and dynamic, having fun and lots of free time, being free from responsibilities etc. They underlined what being young looked like in "reality," usually referring to the global neoliberal conditions exemplified by the specific social and political context of Turkey. Studying the concept of youth in the context of protest, with many young people who were already members of political groups or organizations, enriched the data considerably in terms of providing input as to how being young related to collective action and the notion of the political.

The subthemes were created to cover young people's perceptions of being young without overlooking the complexity of the function of the notion of youth in social life and the heterogeneity of the participants' experiences of youth. The detailed analysis of this theme initially touches upon the points through which young people perceive being young in terms of capabilities and possibilities. Regarding the "downside" of being young, the interviewees' feelings of oppression and exclusion were studied by focusing on the concepts of generational conflict, social pressure, and gender. In addition, anxiety about the future, as a widespread and overarching concern for young people, was investigated by paying special attention to the phases during which it gradually manifested itself in the participants' experiences, usually starting with the university entrance exams and continuing up to the post-graduation period. It was striking to see that, in some cases, depending on the form and extent of their political engagement, young people primarily framed youth—and therefore themselves—as political actors. Accordingly, the research explored being young as a potential for political action on the one hand, and on the other hand the limitations faced by young people taking political action. Lastly, it was studied how ordinary modes of being young were turned into exceptional ones in the extraordinary atmosphere of the Gezi protests. This last part disclosed that the politics of the Gezi protests could hardly be understood without looking closer at its transformational effect in terms of the experience of being young.

Young people's relationship with politics and their notions of the political first emerged as a theme in the literature review stage of the project and then became the primary focus of this study. With a critical eye on the literature on youth political participation, in which young people are mostly depicted through dichotomies such as apolitical-political, this work sets out to study youth and politics beyond these dichotomies. The unexpected, visible mass participation of supposedly apolitical young people in the Gezi protests had already proved that the political-apolitical binary is neither valid nor reliable when it comes to explaining young people's relationship with politics in Turkey. Inspired by current theoretical discussions on the changing definition of the political, this research widened its focus by looking closer at various spheres of everyday life and various forms of cultural practices instead of spaces and forms that were traditionally considered to be political such as traditional party politics, political organizations, or voting. In addition, disregarding any limited, preexisting definitions, the political as a concept was treated as an open term to be worked on interactively with young people in the course of the interviews. Within this framework, the interviews were seen as open discussions with young people on politics and the political, and thanks to this they yielded a large number of valuable inputs for the analysis. The subthemes of this theme mainly included young people's perceptions of the apolitical youth discourse, their relationships with the traditional political sphere, and their ways of describing as well as practicing the political.

Throughout the interviews, I recognized that, due to having been frequently labeled as apolitical, young people were willing to talk about and criticize the mainstream apolitical youth discourse. Discussing the definition of the apolitical and the function of this concept in their lives, the interviewees also reflected on the discrepancy between the older generations' understanding of the political, and their own, which had resulted in them being perceived as apolitical by their parents and their parents' peers. Moreover, elaborating on the concept of the apolitical in a way offered insights about what the political is and is not according to the young interviewees. Young people's critical perspectives regarding the definition and the spaces of the political usually matched their evaluations of conventional politics in Turkey. The interview data was analyzed by looking closely at the participants' accounts underlining their feelings of distrust and hopelessness regarding the current state of mainstream politics, characterized by a lack of critical thinking, individuality, as well as freedom. Moving on to what really was political for them, young people mostly shared how they thought about, communicated, and practiced the political in their everyday lives. By studying their descriptions, the political was explored by zooming in on individual everyday life experiences in various areas such as social relations, consumption, and local matters. This theme also includes an analysis of young people's ways of responding to issues in these areas by raising awareness, reflecting on values, and expressing themselves in various creative ways.

## 2.4 Reflexive Notes on the Research Process

As mentioned above, reflexivity in this study mainly refers to reflective thinking on the role of the researcher, as well as on the context in which the research was conducted. This section looks at some of the circumstances which have had a significant impact on the research, as well as my own influence on it. Generally, it is important to highlight that the researcher's influence on the research process should not be understood merely as the result of his or her personal traits, affiliations, or socio-economic and political background. Instead, for the sake of deepening the discussion on reflexivity, a broader interpretation should be considered, one that highlights the researcher's methodological choices by specifically drawing attention to the relation between the ontological and epistemological assumptions of the methods used and the research process. In this regard, Natasha S. Mauthner and Andrea Doucet's understanding of reflexivity, which is based on a detailed analysis of their reflexive practices in their own research, warrants attention:

> In particular, we suggest that reflexivity is not confined to issues of social location, theoretical perspective, emotional responses to respondents, and the need to document the research process, aspects of reflexivity which are highlighted within current literature. We demonstrate how more neglected factors such as interpersonal and institutional contexts of research, as well as ontological and epistemological assumptions embedded within data analysis methods and how they are used, can deeply influence research processes and outcomes. (Mauthner and Doucet 2003, p. 418)

Inspired by this broad definition, this study understands reflexivity as a way of relating to the whole research process, where multiple and complex factors influence and shape the form and content of the work. The social position of the researcher—including his or her socio-economic, political, and educational background, as well as gender, age, and ethnicity—affects his or her relationship towards the research participants. The theoretical approach to the research, the formation of the research questions, the main concepts discussed, and the methods used—that is to say, the ontological and epistemological assumptions on which the research is based—made up the lens through which the subjects of the research were observed and analyzed.

Ethnographic interviewing requires establishing a trust-based relationship between the interviewer and the interviewee. Since the interview is organized as a flexible and open-ended conversation, the goal of hearing the perspectives of the interviewees can be achieved only as long as the interviewees feel safe and comfortable sharing their experiences. It is therefore essential to understand at the very beginning the importance of trust in the interviewing process. For this reason, I attempted to be as transparent as possible regarding my personal background before asking interviewees to share theirs, attempting to establish our relationship as a reciprocal one. I introduced myself at the beginning of each interview, giving a detailed account of my educational background and how and why I started this research project. Since most of the interviewees were still university students (both undergraduate and graduate), and many others had just graduated, sharing my educational background did not create a distance between us, but rather worked to establish a common ground on which we were able to start our discussion. Apart from that, if interviewees wanted to know more, such as my place of origin or how I came to Istanbul, I answered, but without giving too many personal details.

Since I was in Germany for my doctoral studies during the protests, I also told my interviewees that I had not been able to witness the protests in person, pointing out that what they were about to tell me about the protests would allow me to imagine their protest experiences and write about them. The interviewees usually told their stories in detail, knowing that I did not have any first-hand experience of the protests, and this permitted me to gather much more and richer data. In addition, not having been a protestor myself also meant that I could take a more "neutral" position and keep a critical distance from the interviewees. Throughout the interviews I tried to maintain my position as the researcher, but also to express empathy.

Since this study focuses on a protest movement, it is necessarily bound up with politics, thus the political affiliations and/or sympathies of both the researcher and the interviewees, the interaction between them, and the political atmosphere at the time affected the content and the dynamics of the conversations. Maintaining my position as a researcher without expressing attachment to any political role or affiliation served to limit the effects of potential differences of political opinion in the interviewing process. When I was asked to share my thoughts on a political issue in the interviews, I did not abstain from expressing my viewpoint, and this strengthened the feeling of trust between the interviewees and myself.

Reflexivity also fosters awareness of the effects of the research process on the participants. For most of the young interviewees the interviews were self-reflective experiences through which they revisited the protests, reminded themselves of the important events from that period, and thought about their lives in relation to their participation in the protests. This had different effects on different participants. The interviews conducted in the first phase of the fieldwork especially indicated that most participants were still hopeful that the transformation started by the protests would continue. Remembering their experiences of solidarity, collective spirit, and alternative modes of living (such as communal living in the park) therefore mostly seemed to strengthen their hope, and belief in change. On the other hand, the interviewees I talked to in the second part of the fieldwork frequently expressed disappointment, since many things they had hoped would change and improve after the protests did not change at all, and some even deteriorated further.

Some participants recalled traumatic moments from the protests, such as being caught in a police attack, being injured, witnessing a close friend or relative being injured in the clashes between the police and the protestors. In such cases the content and flow of the interview changed dramatically. I proposed pausing the recording or taking a break if the interviewee seemed uncomfortable. If the interviewee wanted to continue, I let him or her decide when and how to change the subject, and then I continued with new questions.

In addition, individual participants' modes of relating to the research and their reactions to being interviewed were also considerably different from each other. In their responses to the information on the ethical rules and considerations of the research, some were more skeptical and anxious regarding the protection of their personal information. In those cases, I reassured them that the rules of anonymization and absolute confidentiality would be respected, and in addition offered to keep in touch with them if they had any other questions regarding ethical issues. Some interviewees, by contrast, seemed comfortable sharing their thoughts and experiences from the protests, and were not concerned as to whether their information would be kept anonymous or not. One interviewee's account is telling in that sense, as he states that everything he says can be shared openly. The whole story is "legitimate," and there is no need to hide any details:

> What is said here can be shared because everything here is legitimate. Actually, everyone approached this issue with similar concerns as you; I mean, as if the Gezi protests were something illegal or illegitimate. It was totally legitimate… It was a movement, which was our basic right. For that reason, it is good that it is being written down, it is a piece of historical heritage which will be left by us to future generations. (30, employed, leftist activist and football fan)

The fact that I was studying an academic "hot topic" also had to be taken into consideration in the research process. The Gezi protests became a popular topic for discussion and research soon after they took place in May and June 2013; academic interest both in Turkey and internationally prompted comments, conferences, workshops, and new research projects on many aspects of the protests. Studying a new and developing field requires a dynamic and reflexive attitude toward one's own work in terms of its relation to already published works, as well as its original contribution to the discourse. The fact that the protests are a popular academic field of interest affects in various ways the manner in which the research subjects relate to the research. While some young people were excited for their stories to be represented in an academic work, and wanted to go into detail, others were very critical of the researchers studying the Gezi protests, especially because of their way of approaching the research participants as mere sources of information. One interviewee, who had been interviewed several times after her participation in the protests, was suspicious as to whether she could say anything "original" about the protests:

> We have been objects of so many interviews and academic research since the Gezi protests, it has become such a routine. It has been nearly one year; even in that period, everything was mixed up because of telling the story so many times. I do not know what I can say today. (23, student, LGBT activist)

Some interviewees had had negative experiences with previous interviews. One recounted how what she had assumed to be an academic interview was published in a newspaper without her permission:

> Someone conducted an academic interview with me like you, but then she gave my interview to a newspaper on Facebook. (25, employed, oppositional Muslim)

In these cases, the beginning of the interview was devoted to explaining the aims and priorities of my research in detail, underlining that the interviewee's perspective was important, and that there was no need for original stories. Furthermore, I underlined several times that the data gathered would not be used without permission.

The data analysis stage of the research also deserves reflexive attention. As explained above, thematic analysis is treated as "a way of seeing". This underscores the fact that the analysis of the data gathered cannot be purely objective or neutral; rather, it is what the researcher sees and how she or he reads the data. It is therefore important to state that my analysis of the field material is unique to this work. Undoubtedly, there are various other possible ways of gathering, studying, and analyzing data on the same topic. However, this does not compromise the scientific quality of the research; on the contrary, being aware of one's own role in shaping the data and its analysis stresses the uniqueness of ethnography as methodology.

> The advantage of ethnography as a research tool then becomes clear: thanks to its emphasis on context and self-reflexivity, ethnography explains the viewpoint from which the researcher makes her interpretation. (Bray 2008, pp. 314–315)

The data analysis in this work is thus the ethnographic recording of thematically organized interpretations of the data gathered by the researcher.

## 2.5 Limitations of the Fieldwork

The previous sections draw a detailed picture of the methodological approach and the fieldwork process. In addition to this, it is also important to note the limitations of the methods used and discuss advantages and disadvantages of the methodological approach employed in this study. Moreover, it is also crucial to point out potential new discussions regarding the methodological foundations of the research in relation to its epistemology and ontology.

First of all, as discussed above, this ethnographic fieldwork does not claim to be representative of the large and diverse groups of young protestors in the Gezi protests. Rather, it focuses on individual experiences. Limiting the number of participants to 40 was a conscious and deliberate choice. The material emerging from this limited number of interviews provided very rich data, in sufficient quantities. However, it is still important to note that this research is limited in terms of both the number and the diversity of its participants. Even when an effort was indeed made to ensure the diversity of participants, especially in terms of their political background, the number of people from specific backgrounds I reached was limited. For example, increasing the number of young people with no political affiliation could further contribute to the conceptual discussions on the notion of the apolitical as well as on the various (thought to be) non-political ways the political is experienced in everyday life. Similarly, more interviews with practicing Muslim protest participants would have increased the share of critical discussion of the experience of collective identity in the protests as well as prompted exploration of different facets of being young in Turkey by focusing on religious affiliation as a determinant.

In-depth interviews were selected as the main research tool because they provide unique and rich data that cannot be gathered through other methods. However, since the methods used shape the form and content of the data collected to a great extent, the field material gathered in this study is limited. Using other qualitative methods in addition to in-depth interviews would certainly have yielded larger quantities of more diverse data. For example, an extended participant observation phase, coupled with focus groups for young people to discuss their notions of the political would have probably provided valuable insights into young people's experiences of being young and everyday political practices.

Lastly, studying a recent and complex phenomenon in a dynamic, rapidly changing socio-political context comes with both advantages and disadvantages. On the one hand, studying the protests from the perspective of young protestors opens up a new way of understanding young people and their relation to politics contextualized within the current dynamics, and therefore offers new insights into the lives of young people in Turkey today. On the other hand, it is difficult to gather up-to-date information in a context in which young people are responding to rapidly developing political and social events. It is thus important to stress that, in treating its topics and research questions, this book is limited by the overall context and duration of the research period. It takes a snapshot of young people and their relationship with politics by focusing on their participation in the Gezi protests, and presents a history of youth and politics, whose dynamics will obviously change over time.

## References

Baiocchi, G., & Connor, B. T. (2008). The *Ethnos* in the Polis: Political ethnography as a mode of inquiry. *Sociology Compass, 2*(1), 139–155.

Blee, K. M., & Taylor, V. (2002). Semi-structured interviewing in social movement research. In B. Klandermans & S. Staggenborg (Eds.), *Methods of social movement research* (pp. 92–117). Minneapolis and London: University of Minnesota Press.

Boyatzis, R. E. (1998). *Transforming qualitative information: Thematic analysis and code development*. Thousan Oaks, London, New Delhi: Sage Publications.

Ercan Bilgiç, E. & Kafkaslı, Z. (2013). *Gencim, Özgürlükçüyüm, Ne İstiyorum? #direngeziparkı Anketi Sonuç Raporu*. Bilgi Üniversitesi Yayınları.

Bray, Z. (2008). Ethnographic approaches. In D. della Porta, & M. Keating (Eds.), *Approaches and methodologies in social science: A pluralist perspective* (pp. 296–315). Cambridge: Cambridge University Press.

Della Porta, D., & Keating, M. (2008). How many approaches in the social sciences: An epistemological introduction. In D. della Porta, & M. Keating (Eds.), *Approaches and methodologies in social science: A pluralist perspective* (pp. 19–39). Cambridge: Cambridge University Press.

Denzin, N. K., & Lincoln, Y. S. (2011). Introduction: The discipline and practice of qualitative research. In N. K. Denzin & Y. S. Lincoln (Eds.), *The Sage Handbook of Qualitative Research* (4th edition) (pp. 1–26). Los Angeles, London, New Delhi, Singapore, Washington DC: Sage Publications.

Erdoğan, E. (2015). Siyasal Psikoloji Siyasal Katılım Hakkında Ne Öğretebilir? Gezi Protestoları'na Katılanlar Üzerinden Bir Değerlendirme. *Marmara Üniversitesi Siyasal Bilimler Dergisi, 3*(5), 31–58.

Flick, U., von Kardorff, E., & Steinke, I. (2004). What is qualitative research? An introduction to the field. In U. Flick, E. von Kardorff, & I. Steinke (Eds.), *A Companion to Qualitative Research* (Bryan Jenner, Trans.) (pp. 3–11). London, Thousand Oaks, New Delhi: Sage Publications. (Original work published 2000)

Gümüş, P., & Yılmaz, V. (2015). Where Did Gezi come from? Exploring the links between youth political activism before and during the Gezi protests. In I. David & K. Toktamış (Eds.), *Everywhere Taksim: Sowing the seeds for a new Turkey at Gezi* (pp. 185–200). Amsterdam University Press.

Hammersley, M., & Atkinson, P. (2007). *Ethnography – Principles in practice.* London: Routledge.

KONDA. (5. June. 2014). *Gezi Report: Public Perception of the "Gezi Protests", Who were the people at Gezi Park?. Retrieved January 5, 2022, from*https:/konda.com.tr/wp-content/uploads/2017/03/KONDA_Gezi_Report.pdf.

Mason, J. (2002). *Qualitative Researching* (2nd Edition). London: Sage Publications.

Mauthner, N. S., & Doucet, A. (2003). Reflexive accounts and accounts of reflexivity in qualitative data analysis. *Sociology, 37*(3), 413–431.

Özdüzen, Ö. (2017). Cinema-going during the Gezi protests: Claiming the right to the Emek movie theatre and Gezi Park. *Social & Cultural Geography, 19*(8), 1028–1052.

Papazian, H. (2016). Between Gezi Park and Kamp Armen: The intersectional activism of leftist Armenian youths in Istanbul. *Turkish Studies, 18*(1), 56–76.

Ryan, G. W., & Bernard, H. R. (2003). Data management and analysis methods. In N. K. Denzin & Y. S. Lincoln (Eds.), *Collecting and interpreting qualitative materials* (2nd edition) (pp. 259–309). Thousand Oaks, Calif.: Sage Publications.

Tanyaş, B. (2015). Gençler ve Politik Katılım: Gezi Parkı Eylemleri'nde 'Apolitik' Nesil. *Eleştirel Psikoloji Bülteni, 6,* 25–50.

Türkmen, B. (2014). Gezi Direnişi ve Kadın Özneler. *Kültür ve İletişim, 34,* 11–35.

Uyan-Semerci, P. (2013). Nitel Yöntem Nedir? Ne Değildir?: Sosyal Bilimlerde Derinlemesine Analizin Gerekliliği. In P. Uyan-Semerci & E. Erdoğan (Eds.), *Farklı Pencereler Farklı Manzaralar: Sosyal Bilimlerde Yöntem Tartışmaları* (pp. 87–106). Istanbul: Hiperlink Yayınları.

Yavuz, O. (2014). Kahrolsun Bağzı Veriler. In V. S. Öğütle & E. Göker (Eds.), *Gezi ve Sosyoloji: Nesneyle Yüzleşmek, Nesneyi Kurmak* (pp. 99–123). Istanbul: Ayrıntı Yayınları.

# Part I
# Youth in Collective Action

# 3 Studying Youth and Social Movements

The present chapter elucidates this work's approach in studying the Gezi protests as a social movement, with detailed discussion of how this research handles the protests with reference to current discussions as well as the theoretical toolkit existing in the recently developing social movement's literature. In other words, the conceptual framework and the main literature employed as a background for the study of the Gezi protests and young participants' practices within this process are clarified. This work does not merely examine a youth movement, but rather focuses on young people in a wider protest environment including various other participants of different ages and with different socio-biographical features. Therefore, in addition to understanding young people's concerns and lifeworld's linked with their motivations for participating in the Gezi protests, deciphering the links and relations between young people's actions and the dynamics of the broader context both in terms of the social movement they are acting in as well as the global scenery of collective action is of great importance. Accordingly, this chapter also attempts to situate young people's collective action in the Gezi protests within its wider circumstances.

The first part of the chapter explores different theoretical approaches to studying collective action in recent decades. On the basis of this review, it will be discussed why this work situates itself closer to a cultural approach, and how the performative perspective inspired the interpretation of the protests and the protesters' experiences in this research. The importance of understanding individual agency, and also the process of the formation of collective identity in the context of social movements, are then examined. The second part of the chapter brings into focus youth in social movements. To this end, first, important instances of youth collective actions in Turkey from the 1960s to the 2000s in Turkey are covered. Second, the literature on the recent global wave of protests between 2009 and 2013, within which the Gezi protests are also situated, is reviewed,

 
P. Gümüş Mantu, *Redefining the Political: Youth Experiences of Collective Action in Turkey*, Kulturelle Figurationen: Artefakte, Praktiken, Fiktionen,
https://doi.org/10.1007/978-3-658-40565-6_3

especially with regard to the interpretations of young people's participation in those mobilizations. Lastly, the characteristics of the Gezi protests in the way that they have been analyzed as a social movement and how especially the young protesters' role and presence throughout these protests have been studied are discussed. In this way, this chapter builds the theoretical background upon which the next chapter—Chap. 4—is based while interpreting young people's experiences of the Gezi protests. Thus, these two chapters together are intended to provide a comprehensive and thorough analysis of youth in the Gezi protests, informed by the theoretical discussions on the study of social movements and situated within the recent relevant literature on young people in social movements both in local and global instances.

## 3.1 Social Movements

It would not be wrong to describe the field of social movement research as an emergent one. Studies on social movements date back to theories derived from psychology which take a reductionist view according to which people's collective action is merely "the manifestation of feelings of deprivation experienced by individuals in relation to other social subjects, and of feelings of aggression resulting from a wide range of frustrated expectations" (della Porta and Diani 2006, p. 7). From this perspective, collective action is mostly thought to be a reaction to critical situations caused by inconveniences or gaps in the political and economic structure, or linked to disruptive psychology.

This way of analyzing collective action, as a reflection of problems in the social order, began to lose its validity with the emergence of studies from the discipline of sociology, which focused on collective behavior. Moving the focus from collective psychology to collective behavior, this approach highlighted the observable actions of individuals/groups. Moreover, collective behavior has been described as "behavior concerned with change, and social movements as both an integral part of the normal functioning of the society and the expression of a wider process of transformation" (della Porta and Diani 2006, p. 12). Thus the notion of social movement has been reframed not as a problem to be fixed but an important part of the social reality and human action, which is strongly linked with social change and transformation, and therefore worth being discussed and understood in detail.

Two main points of criticism have been leveled against the collective behavior perspective: overlooking the organizational processes of social movements and being limited to describing the structural conditions in which movements emerge. These critiques prompted the development of approaches which focus on organizational strategies devised by rational and strategic actors and the structural origins of social conflicts respectively (della Porta and Diani 2006, p. 14). It is no coincidence that the rise of organizational-structural approaches to studying social movements in the early 70s concurs with the rise of new social movements. Many sociologists who were involved in studying this field at the time were themselves active in or sympathetic to the movements that constituted their subject matter (Goodwin, Jasper and Polletta 2001, p. 4). Simultaneously, being the actors of the movements that they focused on, these sociologists fostered a perspective that highlights the inner and outer resources of the organizational processes of movements and also the specific contexts in which they are built. Therefore, their participation in the field was related to the growing interest in the "'how' of organization-building, strategy, and tactics than in the 'why' of motivation" (Goodwin, Jasper and Polletta 2001, p. 5).

These structural explanations of people's collective action widely known as theories of resource mobilization and historical systemic conflicts have been dominant in the field for a long time. Moreover, these perspectives have informed the development of the grand theories of social movements and have therefore contributed to the emergence of social movement studies as a subfield in sociology. Along with this trend, some examples of prominent themes of social movement research have been: internal and external resources of collective action, goals and rational decision processes, social and political context in which movements emerge, changing forms of social conflict from class to identity and symbolic power, actors' power/ability to influence the political sphere, and also the success and failure of people's collective action. Through these themes, structural approaches underline people's ability to organize, the conditions in which they do so, and their demands for transformation in relation to the context in which they are living. However, this emphasis on the structural factors was limited in another way, namely in terms of analyzing the micro-processes, the individual, and more specifically the manners and motives of participants' involvement in movements, which are crucially important for gaining a deeper understanding of a social movement. These under-researched themes were then adopted as the center of research by the cultural approach, to which this study more closely adheres within the existing literature on studying social movements. At this point, it is possible to summarize that research in the field has developed mainly in two opposite directions: as the dominant and structural approach on the one hand and the cultural and constructionist tradition on the other. While the former highlights economic resources, political structures, formal organizations, and social networks, the latter is centered on frames, identities, meanings and emotions (Goodwin and Jasper 2004, p. vii).

### 3.1.1 Cultural Approach: Thoughts, Emotions, Morals, and the Performative Perspective

Regarding the limits of existing theories in analyzing social movements, James M. Jasper suggests studying social movements by focusing on culture. Culture occupies the central place in understanding collective action, depending on the definition of the concept as such:

> Culture is composed of shared thoughts, feelings, and morals, along with the physical embodiments we create to express or shape them. It is through cultural processes—from singing to reading to marching together down a street—that we give the world meaning, that we understand ourselves and others. (Jasper 2014, p. 7)

In this definition, Jasper lists the most important concepts of the cultural analysis of social movements, such as individuals/groups' thoughts, feelings, and morals, which are all shared and also expressed and shaped by physical embodiments. However, what is most striking about this definition is that Jasper characterizes social movement as a cultural process through which people give meaning to the world around them as well as understand themselves and others. Therefore, looking closer at a social movement is also a way of looking closer at how people see, experience, and reflect upon the world around them, and, moreover, upon how they relate to themselves and others. In this way, the cultural approach makes it possible for researchers to thoroughly examine and understand how individuals take part in collective action, without being limited to psychological or behavioral explanations. Inspired by this perspective in studying social movements, this research analyzes the Gezi protests as a cultural process, and intends to discover what this process reveals in terms of young people's perceptions of themselves and their environments as well as their relations to the socio-political context they were living in.

As quoted above, Jasper states that thoughts, feelings, and morals are the main components of culture. Thoughts are usually taken as the informative data of people's cognitive processes linked to their protest participation and existence in social movements. However, what he rightfully underscores is that feelings and morals are also important and indispensable parts of the cultural process through which researchers can analyze people's ways of relating to the social movements they are acting in.

By way of a critical review of the ways in which emotions have been handled in social sciences, Jasper calls for going beyond the historically established 'emotional-rational' duality in order to be able to understand the function and meanings of emotions in people's actions. He emphasizes that feeling and thinking are not separate, but rather "parallel, interacting processes of evaluating and interacting with our worlds" (Jasper 2011, p. 14. 2). Emotions have long been absent in the field of social movements; especially the historical structural approaches consciously left emotions aside in search of a more 'objective and realistic' analysis of social movements. This very ignorance was itself an important characteristic of these dominant approaches in social movements; in parallel to what was mentioned above, it was mostly an objection to the psychological understanding of collective action. Craig Calhoun, in his discussion on emotions in the study of social movements, writes how emotions drifted in and out of the field through decades by different approaches as such:

> With the bathwater of some very serious biases, the baby of emotions was commonly thrown out. It is hard to get emotions back into the field partly because they were not merely neutrally absent from it but expelled in an intellectual rebellion that helped to give the field its definition. (Calhoun 2001, p. 48)

Taking emotions as one of the main parts of the cultural approach in studying social movements requires a holistic understanding of individuals' experiences, which means both thinking and feeling, or thoughts and emotions, are considered as dynamic relational parts of a whole. This way of thinking, emphasizing the relatedness of the components of the cultural process, does not hold emotionality to be an additional aspect, but investigates how emotions are at play within thoughts as well as morals. Therefore, this perspective comes with an understanding of the social including and through emotions, in other words, a process of thinking that focuses on the sociology of emotions:

> Any serious sociology of emotions must be more than an ad hoc call to look at the additional variable of 'emotionality.' It requires frameworks for bringing intrapsychic and cultural dimensions of meaning and action into clear relationship with social organization (Calhoun 2001, p. 51).

Following this sociological perspective, a cultural analysis of social movements should include emotions not in an eclectic but in a relational manner by paying special attention to how emotions affect and are affected by the other aspects of the cultural processes. In this way, it becomes possible to discover in which specific ways emotions create and shape collective action. More specifically, emotions are everywhere in a protest, "they motivate individuals, are generated in crowds, are expressed rhetorically and shape stated and unstated goals of social movements" (Jasper 2011, p. 14.2). Thus the present study analyzes the Gezi protests by giving special consideration to emotions while focusing on young people's accounts of their participation in the protests. How feelings of unhappiness, oppression, and anger were felt immediately before the protests and how these feelings transformed into feelings of happiness, solidarity, and trust as the constituents of a collective identity will be studied in detail in the next chapter.

Along with emotions, participation in a collective action or being a member of a social movement is strongly linked to sentiments about what is right or wrong for oneself, in other words, to morality. Moral perceptions, in reference to which one feels something is wrong, usually mark the point by which people are directed towards politics (Jasper 2014, p. 8).

> When we learn or experience something that suggests the world is not morally as it seems, our indignation has a strong visceral aspect. We feel betrayed by the world as well as by other people (Jasper 2010, p. 88)

Moreover, too often protest movements escalate triggered by "moral shocks", such as a violent repression of a peaceful protest (Jasper 2011, p. 14.8). Feelings of disappointment resulting from ethical considerations regarding the way people are treated within the given socio-political context mostly act as the ground upon which protest movements are built. Regardless of their ideological, socio-economic, and other differences, moral visions create the potential for crowds of people to come together on the basis of shared humanistic values and perceptions of right and wrong. The violent repression of a peaceful protest of environmental activists against the destruction of the trees in the Gezi Park is an example of an event that caused moral shock. This moral shock started the Gezi protests that spread all over the country, and eventually turned into one of the biggest protest events in the history of Turkey. Therefore, how young protesters went through this moral shock and how their ethical considerations shaped their participation in the protests is analyzed in the following chapter as part of the analysis of the field data. By looking closer at how morality played a role in the protests for young people, an attempt is made to establish a link to the discussion about how values act as the constituents of the protesters' notions of the political, which is examined in detail in Chap. 8.

How we create, express, and share the main components of culture, as thoughts, feelings, and morals, more precisely, how we perform them, is also crucially important. By employing a performance theory informed approach, it becomes possible to understand how participants act through the cultural processes of collective action and "how social movements move" (Eyerman 2006):

> …the application of a theory of performance calls attention to the place and space of movement, as well as how opposition is performed. Performance theory focuses on corporality, presence, and the pre-discursive, while at the same time including it. This allows us to better address questions concerning what happens when people enter a movement, how this affects their actions and the actions of the others, and to ask how social movements move. (Eyerman 2006, p. 193)

Therefore, performance theory allows a more nuanced and deeper understanding of social movements. By applying a performative perspective to the study of social movements, it becomes possible to shed light on the link between the cognitive background and the acts/actions of the participants in a protest. The move from thinking (and feelings and morals) to taking action occurs in two interconnected layers: one in which framed emotions transform into action, and the other in which individually based, diffused experiences turn into focused and collective ones (Eyerman 2006, p. 196).

Moreover, the cultural-performative approach also points to the reciprocal relationship between social movements and society. Social movements are not only affected by the context of the broader culture, they also affect the conditions that surrounded them. Therefore, discussions on analyzing social movements revolve not only around the moves and processes inside the movements but also around how the movement is in a continuous dynamic relationship with the larger context:

> A performative view of culture stresses that social movements are not just shaped by culture, they also shape and reshape it. Symbols, values, meanings, icons, and beliefs are adapted and molded to suit the movement's aims and frequently are injected into the broader culture via institutionalization and routinization. (Johnston and Klandermans 1995, p. 9)

Young individuals' processes of being spurred into the action in the Gezi protests as well as being a part of the collective Gezi Spirit are studied in this work following the cultural approach informed by the performative perspective. The use of this approach also brings about a special emphasis on the interaction between the cultural dynamics of the Gezi protests and the larger socio-political and cultural atmosphere of the country.

### 3.1.2 Individuals and Collective Identity

Along with the framework discussed above, young people's experiences of the Gezi protests are studied in this research project as a whole regarding their thoughts, emotions, and moral views. Moreover, in accordance with the cultural approach's emphasis on the individual, young participants are focused on as individuals primarily. In this way, each and every young person's story included in the field research can be discovered and interpreted in detail, and therefore the differences of their experiences are not reduced to a single archetype of "the protester", but have been analyzed in a way that acknowledges their diversity. As stated by Manuel Castells who studies recent online and urban-based protest movements:

> […] While grouping people's experiences into analytical categories of social structure is a useful method, the actual practices that allow social movements to rise and change institutions and, ultimately, social structure, are enacted by individuals: persons in their material flesh and minds (Castells 2012, p. 13).

Researchers thus have to understand the motivation of each individual in detail. However, how these individuals connect to each other and create a collective identity is also one of the most important topics of analysis in this research. The "individual vs. collective" discussions are also reflected in the structural-cultural tension in literature on social movement studies. While searching for a new way of studying collective identity, Alberto Melucci critically points to the individual-collective duality in the field:

> The study of social movements has always been divided by the dualistic legacy of structural analysis as a precondition for collective action and the analysis of individualistic motivations. These parallel and sometimes intertwined sets of explanations never fill the gap between behavior and meaning, between 'objective' conditions and 'subjective' motives and orientations (Melucci 1995, p. 42).

Thus, encouraged by Melucci's well-grounded critique, this research studies individuals and the collective identity in relation to each other by looking closer to the links between participants' worlds of meaning and their actions, as well as their "subjective" positioning and their "objective" circumstances. In doing so, following Melucci, a processual approach is employed for understanding the individual in the creation of collective identity:

> Collective identity is an interactive and shared definition produced by several individuals (or groups at a more complex level) and concerned with the orientations of action and the field of opportunities and constraints in which the action takes place. By 'interactive and shared' I mean a definition that must be conceived as a process because it is constructed and negotiated through a repeated activation of relationships that link individuals (or groups). (Melucci 1995, p. 44)

Thus, Melucci underlines the active roles of the individuals in the dynamic construction of the collective identity and places special emphasis on the relations between individuals. Within this framework, understanding how individuals decide and continue to act in collective bodies is just possible through employing a processual research practice, which focuses on how individuals exist throughout the changing conditions and context of a movement. This also allows the diverse, multifaceted, and unstable experiences within a movement to be discovered and it therefore becomes possible to question "the existence of a movement as a homogenous actor" (Melucci 1995, p. 55). Following Melucci's approach, the concept of collective identity will be addressed here as a work-in-progress and heterogeneous entity which were created by individuals within the complex dynamics of the course of the protests. The notion of the 'Gezi Spirit', in other words the configuration of the collective identity formed in the course of the protests, will be specifically analyzed in order to decipher it in terms of the process through which it was created by the individuals/groups involved in the Gezi protests.

In light of the above-mentioned discussions, this work studies the young participants of the Gezi protests as individuals, with their thoughts, emotions, and moral values, and looks closer at the dynamics and features of the processes through which they became protesters as well as components of the collective identity built throughout the protests. The diversity of the ways in which young people experienced these processes has been paid attention in order to uncover the heterogeneous character of the Gezi protests and develop an in-depth and process-based understanding of them, informed by the research participants and analyzed from a cultural-performative perspective in studying social movements.

## 3.2 Youth and Social Movements

The recent global wave of protests from 2011 on, and the visibility of young people as one of the main agents of these uprisings have resulted in an upsurge of discussions on youth and social movements. Doubtlessly, these movements were not the first time young people took a stand for social and political change as a collective. Even if a detailed historical overview of youth movements is beyond the limits and the focus of this study, it is relevant in this chapter to take a closer look at how young people in social movements, particularly in Turkey, have been studied within the specific circumstances over different periods. Looking closer at the youth phenomenon in the field of social movements in Turkey throughout the recent decades contributes to the analysis of the discussions on the collective agency of the young participants of the Gezi protests. Additionally, examining the various cases and spaces of youth collective action in recent years in Turkey will provide insights for understanding the roots of the unexpected coalescence of young people from different backgrounds under the banner of the collective spirit of the Gezi protests, which is characterized by diversity and solidarity.

Prior to turning the focus to young people in social movements in Turkey, it would be worthwhile to highlight some of the crucial periods characterized by young people's collective action on a global scale. Doing so will shed light on the possible links between the emergence of local/national youth movements and their global counterparts. In this sense, it is worth paying attention to the spread of student uprisings starting in the second half of the 60s and peaking with Paris '68, and, similarly, the rise of youth countercultures as oppositional social movements and, more recently, the anti/alternative-globalization movements in the 90s and 2000s, in which young people played a prominent role.

Starting with the Berkeley Free Speech Movement in 1964 at the University of California at Berkeley, student movements continued to arise across the United States as well as in Europe, and they also inspired upheavals in Africa, Latin America, and Asia (Sukarieh and Tannock 2015, pp. 92–93). Thus, disseminated worldwide, the global student movement of the 1960s was a milestone in the history of youth collective action, especially with regard to its impacts on the form and content of successor movements as well as long-term political and cultural influence. Having origins in the civil rights protests which began in the USA in 1950s and the peace marches and demonstrations in Europe, these movements had anti-militarist and anti-imperialist agendas and adopted a wide range of issues, such as second wave feminism, environmentalism and global justice (Roberts 2015, p. 954; Sukarieh and Tannock 2015, p. 100). With these protests, the university campus emerged as a space of social, political, and ideological struggle and, moreover, "the idea that young students can be and often are serious and legitimate political actors in society was firmly established" (Sukarieh and Tannock 2015, p. 99). This era also gave distinction to the notion of youth countercultures, within which young people's collective cultural practices are also analyzed as important aspects of youth movements. The 60s and 70s countercultures' disaffiliation was described as being directed against the middle-class culture in which their members had grown up and accordingly they were defined as mostly ideological and cultural. Young people in counter-cultural movements of that period "directed their attack mainly against those institutions which reproduce the dominant cultural-ideological relations—the family, education, the media, marriage, the sexual division of labor" (Clarke et al. 2006, p. 62). Therefore, over the years, youth countercultures have been manifestations of how cultural practices, particularly lifestyle, everyday life relations, music and arts, as well as aims and expectations regarding the future have served as fields of transformation for young people. In other words, countercultural movements revealed that the revolutionary practices in the cultural field, which seem to be limited to the personal sphere, were intertwined with and even at the heart of the political.

The end of the 90s witnessed a global wave of anti-globalization mobilizations, especially targeting the policies of global corporate bodies such as the World Trade Organization, the International Monetary Fund, the World Bank, and the G8 Summit. Beginning in 1999 in Seattle and continuing in the following decade, numerous protests against the meetings of those international organizations were organized. Moreover, social forums were founded as places for building global networks alternative to those run by global elites. This cycle of movements was another important period of collective action in which young people had a visible role. Alter-activism, as an emerging culture of political participation among young participants of these anti-corporate and alternative globalization movements, is defined by referring to an ethics of openness, local-global networking, and organizing across diversity and difference. Furthermore, creative forms of action and an emphasis on experimentation have been underlined as the characteristics of the young alter-activists' discourses and practices (Juris and Pleyers 2009, p. 63). It can thus be said that anti-globalization protests have been events in which young people have developed new repertoires of action and discovered new ways of acting as a collective.

### 3.2.1 Youth Collective Action in Turkey: From the 1960s to the Post-1980 Period

Young people's autonomous and oppositional collective action in Turkey, being more readily comparable to the recent examples of mobilizations such as the Gezi protests in which young people played a prominent role, also took place in the 1960s, parallel to its global counterparts at the time.[1] In particular, the years after the military coup of May 27, 1961[2] were a period in which youth movements were remarkably frequent and prominent. The libertarian environment achieved by the broadening of the basic rights of freedom of expression and organization in the new 1961 constitution after the coup, as well as the widespread influence of the European and US youth movements in 1960s, have influenced the upsurge of youth movements in Turkey. As the new constitution made it possible for young people to express leftist ideas, they were soon faced with accusations of being communists (Kabacalı 2007, p. 159). After 1965, students significantly increased their organizational capacity by founding new organizations such as Opinion Clubs, which were more autonomous and leftist institutions compared to

---

[1] The previous periods undoubtedly also witnessed young people's collective presence in social movements in Turkey. The emergence of youth as a political actor in the history of modern Turkey dates back to the early 20th-century Ottoman Empire, to the Young Turk Movement's appearance as a revolutionary group postulating a transformation of the political and social system (Lüküslü 2005, p. 31). Following this, the social and political significance of youth has also been acknowledged with regards to the struggle for national independence as well as the country's republican periods (Lüküslü 2005, Neyzi 2001). However, in these instances, young people were either acting as the missionaries of the regime or collaborating with other political agents in order to affect the social and political situation of the country, and in that sense do not seem to be the predecessors to the later youth movements emerging from the 1960s on. Therefore, they are not included in the analysis in this part of the chapter.

[2] The military coup that took place on May 27, 1961, was mainly a result of and response to the dramatic change through which the country had been going following the end of World War II. With the transition from one-party rule (under the Republican People's Party) to a multi-party system in 1946, the Democrat Party, mainly based on rural support, overturned the government of Republican People's Party in 1950. The Democrat Party, which is described by Şerif Mardin as "representing the aspirations of persons who desired a society more open to capitalism and also to responsible parliamentary government," had a downgrading effect on the class of Turkish officials, especially through a decrease in their economic power and prestige (Mardin 1978, p. 233). Toward the end of the 1950s, bureaucracy and intellectuals witnessed a dramatic transformation in values of the society while students grew highly critical about the policies of the government (Kabacalı 2007, p. 109). As the party in power was charged with corruption and authoritarianism (Neyzi 2001, p. 418), this period came to an end with the military coup on 27 May 1961, for which student groups were very influential in paving the way (Mardin 1978, p. 233).

the previously established semi-corporatist ones (Alper 2010, p. 81). This period then witnessed a widespread politicization of youth during which young people gradually became more divided as the two opposed camps of "the leftists" and "the rightists" (Neyzi 2001, p. 419). Yet it is important to note that the most important student movements in that period were anti-governmental and left-wing student mobilizations (Mardin 1978, p. 231).

The massive explosive spread of students' boycotts and campus occupations in Turkish universities in June 1968 demanding university reforms was inspired by their international counterparts, and especially by May '68 in Paris (Alper 2010, p. 83). These student uprisings created a political opportunity by generating a domino effect for other sectors, particularly stimulating the mobilization of workers and peasants as well as teachers. Their method of occupation was adopted by the latter groups and used in factory and land occupations. What is more, students became actively involved in the protests in these different sectors and shared their experiences of organizing protests (Alper 2010, pp. 85–86). Subsequently, toward the end of the 1960s, the student movement gradually began to overstep the borders of legal action. Young people's disillusionment with the status quo and the impacts of parallel movements in Europe, Latin America, and elsewhere coupled with being triggered by groups who took the advantage of the rise of extremism were factors that contributed to the increasing violence that characterized this period (Neyzi 2001, p. 419). The military coup of 12 March 1971 happened on the pretense of wanting to bring this rise of violence and chaotic atmosphere to heel, but in reality it directly hit the leftist youth movements. Several youth movement leaders in their twenties were killed by the regime in that period (Lüküslü 2015, p. 159).[3]

Even if the second half of the 60s marked a peak in terms of the youth movements in Turkey, it can hardly be claimed that it attracted the academic attention it deserves. There exist many biographies, memoirs, and interviews regarding the period, but academic resources focusing on this period are significantly limited and far from diverse (Lüküslü 2015, p. 10). However, an in-depth and well-rounded understanding of youth movements of this period would undoubtedly contribute to a better understanding of the history of youth, the youth collective action, and the relationship between youth and politics in Turkey. The need for a cultural reading of this period seems pressing at this point (Neyzi 2001, p. 419).

[3] Deniz Gezmiş, Yusuf Aslan, and Hüseyin İnan, the three young leaders who were executed in that period, have become symbols of radical leftist struggle in Turkey. A poster of Deniz Gezmiş was hung on the frontage of the Atatürk Culture Centre facing Taksim Square, which became an alternative billboard full of slogans and posters in the course of the Gezi protests.

Demet Lüküslü's work on the youth movement of 1968 in Turkey presents a generational analysis of the period combining historical and sociological perspectives (Lüküslü 2015). Focusing on the characteristics of the '68 generation and contextualizing them in the political and social circumstances of the period as well as analyzing this generations' engagement with literature and especially with humor, Lüküslü insightfully offers an understanding of young people's actions and agency in that period, going beyond official as well as frequently employed revolutionary ideological discourses.

Violence continued to characterize the social and political scene in Turkey until the military coup on September 12, 1980, repressed and silenced the political activities again, violently in many instances. Şerif Mardin, writing in 1978, describes how violence was the defining feature of recent periods, especially regarding the youth in the country:

> Since 1965, Turkish universities have been the stage of violent student demonstrations; since the late sixties guns and dynamite have become part of the campus scene, and in the seventies violent death is an item that figures in the daily life of university students (Mardin 1978, p. 230).

The nationwide politicization of youth associated with the rising violence resulted in young people being perceived as "a threat to the national interest" (Neyzi 2001, p. 419). The previous "vanguards of the state" were made out to be "rebels against the state" in the 60s and 70s. In reality, young people were still loyal to their mission of being the vanguard. However, since they considered the regime illegal, they fought against those in power. Eventually, politicians and the media framed young people as "anarchists" and "terrorists" revolting against the order and creating chaos in the country (Lüküslü 2005, p. 33). The ways young people's collective action was perceived and particularly the notion of "dangerous young people" were not specific to these periods; rather, each political tendency and period seems to create "its own dangerous youth" depending on the social and political context (Lüküslü 2015, p. 195). The perception of young people's politicization by the political authorities in the course of the Gezi protests similarly points to a continuity in terms of the way the youth were framed as a dangerous mass by the official discourse (Lüküslü 2015, pp. 193–195). How these perceptions of youth collective action, as a type of action that poses a threat to society, influence the way young people relate to social movements and politics in general is an important question to ask when studying youth in social movements in Turkey. More specifically, the question as to how these official discourses, which at the same time marginalize and expound the problems of young people's collective mobilizations, were internalized by young people themselves and inherited and transferred over generations needs to be addressed if we are to analyze young people's political agency in various contexts of the different periods.

The military regime, which lasted until 1983, aimed to "rehabilitate" the youth by repressive means (Lüküslü 2005, p. 33), which means that young people who had previously been "overtly" and "dangerously" political were forcefully moved out of the political sphere by the military regime. Even after the military regime ended in 1983 and the civil government took over, restrictions were put in place in terms of the freedom of expression and organization. Based on the new constitution, which was enacted in line with the ideology of the 1980 military coup as well as the mainstream political discourse, these means of devaluing oppositional ideas were effective in keeping society and people's everyday lives at a distance from politics. To put it another way, the monopoly of the officials in the political arena became the norm, "doing politics" was accepted as limited to the parliamentary and party politics, and therefore ordinary people gave up their rights to participate in the decisions and take political action, all of which expectedly resulted in a depoliticization, which most notably influenced the generations that were born and grew up in this political environment. What is important to note specifically here is the establishment of the Council of Higher Education,[4] a product of the 1982 Constitution, as a controlling and regulatory institution above universities that was most influential in keeping politics out of the university campus. From 1983 on, beginning with the government of the center-right Homeland Party (ANAP) led by Turgut Özal, Turkey was introduced to neoliberal policies driven by market ideology, coupled with globalization processes, which made themselves felt in everyday life through increasing consumerism and the proliferation of communication technologies. The post-coup depoliticization went hand in hand with the changing socio-economic structure of the country. These circumstances heavily impacted the social and political milieu in which the post-1980 generation grew up in Turkey. Within this picture, young people of the post-1980 generation have been perceived as "the individualistic consumers socialized in a globalized world" (Lüküslü 2005, p. 34) who are not interested in social and political issues in society. In addition, they have frequently been compared to previous generations in the country:

[4] Yüksek Öğrenim Kurumu (YÖK), for more details, see the official website: http://www.yok.gov.tr/en/web/cohe/history.

> Given the cultural weight of both the Republican and 1968 generations in the public sphere, members of the generation known as the "Özal generation" or the "post-1980 generation" tend to be represented as selfish, individualistic consumers, implying the lack of a sense of collective responsibility. (Neyzi 2001, p. 424)

The young generations' withdrawal from collective action seems to have been the grounds for the youth in the post-1980 period to be judged as an "apolitical" generation. However, the outline of the context described above reveals that the distancing of young people from politics was a consequence of wider social and political changes. Moreover, it seems unjustified to call this a process of depoliticization that is unique/specific to young people. In reality, the post-1980 period is characterized by a widespread phenomenon of abstention from politics for many parts of society. Besides, defining young people as "apolitical" is in itself problematic, as it reveals an overly broad and judgmental way of viewing the youth that ignores the variety of young people's ways of relating to and defining politics.[5]

Even if the post-1980 period seems to be the one in which young people's collective action in politics, especially in terms of their participation in traditional political organizations and street politics, has disappeared from the public eye, it is hardly possible to talk about an absence of young people's political subjectivities in this period. Even if young people's political activities have not been that visible in university campuses especially due to less involvement with traditional leftist organizations or political parties compared to the previous periods, young people have still been active participants of post-1980 social movements that do not necessarily look like youth movements but whose main actors have actually been young people (Kabacalı 2007, p. 241).

The Islamic movement, the Kurdish movement, and the Alevi movement can be considered along the identity-based lines of new social movements which have been on the rise in the post-1980 period.[6] Even if young people have apparently

[5] The problematization of the decrease in the levels of youth political participation in the field of youth studies in general is discussed in detail in Chap. 7 of this book. In addition, the implications of the notion of the apolitical youth in the context of Turkey and young people's interpretations of the apolitical are discussed in detail in Chap. 8.

[6] Undoubtedly, the different and particular dynamics in which each of these three movements emerged in the context of the post-1980 period need to be reviewed in order to be able to understand how and why these movements were on the rise. However, as the focus of this work is on youth collective action, these social movements are just mentioned as the ones through which one can trace young people's participation. For a detailed analysis of Islamism, Alevism, Kurdism, and Feminism as new social movements in Turkey, see: (Şimşek, Sefa. 2004).

been active participants in these movements, there is a remarkable lack—with some exceptions—of literature that analyzes youth in these movements. The Islamic movements' most visible political activities were the demonstrations by female students in which they protested against the headscarf ban on university campuses. Additionally, the campuses were the places where Islamic movements, organized young people in the late 1980s and 1990s (Şimşek 2004, pp. 121–123). Alevi youth, in the context of the Alevi identity revival, were also active in defending and negotiating their civil rights in both NGOs as well as in political parties with social democratic tendencies in the 1990s (Özmen 2011).[7] Scholars focusing on youth and children in the Kurdish movement have emphasized the leading role of Kurdish youth in street politics, as well as various alternative ways of doing politics practiced especially by urban, educated, middle-class young people through the institutionalized Kurdish movement and numerous NGOs (Neyzi and Darıcı 2015, p. 68).[8]

Young women were the main participants of the feminist movement during its rise as an autonomous and reflexive movement (Şimşek 2004, p. 124) in the post-1980 period. In the 1990s, women's studies departments were places where young women came into contact with feminism (Bora and Günal 2002, p. 10). Young feminists also organized themselves in feminist/women's studies student clubs in the following decade. In 2006, "The Forum of Women University Students,"[9] in which 120 female students from 20 universities came together in order to discuss "being a woman in the university" stands as an important example of mobilization of young feminists. After the forum, young women published a manifesto in which they outlined their demands regarding the abolition of gendered regulations and application procedures at the universities, elimination of the barriers to active participation by young women in the decision making processes in the universities, and prevention of gender discrimination and sexual assault on campus.[10] The LGBT movement was formed in the 90s when as part of ongoing struggle against discrimination on the basis of sexual orientation and

[7] For detailed discussion of how Alevi young people relate to and reflect on their Alevi identities and frame their demands accordingly, see: (Ezgin, Özlem. 2013).

[8] For detailed discussion of Kurdish children's political subjectivities, see: (Darıcı, Haydar. 2013).

[9] Üniversiteli Kadınlar Forumu (ÜFK).

[10] The call announced by the forum's organization committee (made up of women from four universities in Istanbul) and the final manifesto were both published, following the event, in the periodical journal of the Boğazçi University Women's Studies Club. For these texts and more on the event, see: *BÜ'de Kadın Gündemi 10,11 [Women's Agenda at BU]*, Spring and Fall 2006, İstanbul: Boğaziçi University.

gender identity, LGBT individuals began to organize in civil society organizations in the country's big cities (Yalçın and Yılmaz, 2014). Despite the lack of data and research on LGBT youth in Turkey, young people's presence in the LGBT movement, either in street protests or civil society activities, has been remarkably prominent.[11] In the 2000s, young people also started to engage in civil society activities focusing on environmental issues, prompted by the realization that they were a generation that was born into and lived through emerging environmental problems at local and global levels (Baykan and Lüküslü, 2010).

In the beginning of the 2000s, young people in Turkey took part in anti-war protests, such as "No To War in Iraq!"[12] as well as anti-globalization protests[13] against NATO, the World Bank and the IMF, and they participated in alternative global networking organizations such as Social Forums (Emiroğlu 2013). The founding of *Genç-Sen* was one of the most important events of the period in young people's collective action—for the first time in the history of Turkey young people organized a union to stand up for their student rights without being associated with any ideologies or political parties (Erdinç 2013, p. 309). Youth social movements have been influenced and shaped by the developments in communication technologies since the beginning of 2000s,[14] similar to their global counterparts.[15] Lüküslü, focusing on three different popular websites ("Sourtimes," "Young Civilians" and "52%")[16] predominantly used by young people in Turkey analyzed how young people expressed their discontentment in

[11] Although the number of studies on young people in the LGBT movement is still severely limited, recent works focusing on life experiences of the LGBT individuals may point to possible topics of discussion for in-depth thinking on the youth in the LGBT movement. For example, Volkan Yılmaz and İpek Göçmen's article on LGBT individuals' experiences of discrimination in domains such as employment, housing, and healthcare reveals how social policies regarding LGBT individuals frequently overlap with youth social policies, and therefore serves as a source of inspiration for a discussion on being young and LGBT in Turkey from a relational perspective (Yılmaz and Göçmen 2016).

[12] For more on anti-war protests in Turkey, see: (Altınay, Ayşe Gül. 2003); (Önen, Yıldız. 2015).

[13] For more on anti-globalization protests in Turkey, see: (Gümrükçü, Selin Bengi. 2010).

[14] For more on young people's participation in online spaces and the changing patterns of youth collective action influenced by digital technologies in Turkey, see: (Telli Aydemir, Aslı. 2011).

[15] For more on young people's changing perceptions of democracy and novel forms of participation in the digital revolutionary age in Middle East and North Africa, see: (Herrera, Linda with Rehab Sakr. 2014).

[16] Turkish titles and websites: Ekşi Sözlük (www.sozluk.sourtimes.org), Genç Siviller (www.gencsiviller.net), Yüzde 52 www.yuzde52.org).

non-traditional, radical, and humorous ways via these online spaces (Lüküslü 2014).

The exploratory review presented here proves the narrowness of academic research focusing on young people in collective action in Turkey. Hence studying them holds potential for gaining altogether new insights, rather than merely adding another work to an already well-researched field. Young people remain one of the undiscovered groups in the history of social movements in Turkey.

As can be seen above, looking closer at youth movements in the recent history of Turkey turns out to be a reading of youth collective action in the context of consecutive military coups, which society mostly experienced through oppressive policies, usually accompanied by violence. Given this background, how young people's collective action and the extraordinary political circumstances frequently ruling the country dialogically influence each other stands as an important field of inquiry regarding youth in social movements in Turkey. Furthermore, how these contextual dynamics shaped society's dominant perceptions of youth collective action and political activity, and also how young people's own understandings of political action, protest, and opposition have been formed within these circumstances deserves analysis with regard to the development of youth collective action across generations. Besides, young people's collective activities in the diverse and rich terrain of social movements in Turkey in the post-1980 period particularly deserve attention with respect to developing an understanding of the sources of youth participation in mobilizations taking place in recent years, i.e. the Gezi protests.

It also seems important to favor a youth perspective for the study of social movements. What are the meanings and implications of being young in the contexts of different social movements in the history of Turkey? What are the limitations and possibilities for young people in movement environments? How do young people act within the age-based social relationships, mostly crystallized as hierarchical mechanisms that exist in collective mobilizations?[17] These sorts of introductory questions focusing on youth as a category and a component of collective action may pave the way to uncovering a great deal, not only regarding young people's agency but also concerning the cultures and structure of social movements in Turkey's history.

[17] Volkan Yılmaz's article on being young in political organizations is an example of studying how age as a social relationship functions in collective movements. For further details, see: (Yılmaz, Volkan. 2010).

### 3.2.2 Youth in the Global Wave of Protests in 2011–2014

The protests that began in 2011 and took place in several different places around the world from the Middle East to Europe and United States have been handled as a new wave of global mobilization (Tejerina et al. 2013) that shares common characteristics despite local and context-based differences. The "Arab Spring" that started in Tunisia and Egypt and later spread through the continent, the Indignados in Spain, protests in other Southern European countries, and the Occupy movements all pointed to similar grievances and analogous collective struggles of people across the world:

> Across North and South, in liberal democracies and repressive autocracies, there have been calls for a change of existing political leaderships, institutional arrangements, and political projects. Protest actions, from mass assemblages and public space occupations to art interventions and digital campaigns, have embodied certain era-defining claims—against austerity and socioeconomic inequality, against the deleterious consequences of market forces, as well as the ongoing struggle for enhanced democracy (Davies et al. 2016, p. 9).

Social inequality, accompanied by global capitalism and contextualized within the current neoliberal conditions, has been emphasized as the common ground on which the new cycle of protests broke out in different locations—in other words, as "the heart" of current global mobilizations (Tejerina et al. 2013, p. 380). What this new cycle of protests made remarkably visible was that in addition to the lower classes, those who were previously well off and satisfied with their circumstances within the system were reacting to the deterioration of their economic and social conditions and anxious about whether they would be able to afford a decent life or not. Moreover, these diverse faces of neoliberal social inequality are not limited to the countries in crisis but are also seen in those characterized by economic growth. Donatella della Porta, on the basis of her analysis of the recent protest waves, insightfully suggests that the mobilizations of 2011 revealed the pauperization of the lower classes as well as the proletarization of the middle classes and the degree of social inequality that emerged as a result of neoliberalism in cases of either economic growth or decline (della Porta 2017, p. 17). The pivotal role of inequality as the main force driving the recent mobilizations is argued to necessitate "rethinking previous decades of New Social Movements analyses that focused mainly, or solely, on issues of culture and collective identity" (Tejerina et al. 2013, p. 381). Analyzing anti-austerity protests as reactions to a legitimacy crisis of the late neoliberal system, with reference to the conceptualization of "crisis of legitimacy" by Habermas, della

Porta focuses on how changing political conditions and specific developments in capitalism are interrelated. Therefore, "in order to understand today's movements in times of socioeconomic challenges, we clearly need to bring capitalism back into analysis" (della Porta 2015, p. 9).

Researchers who study the social background of this new wave of protests often employ Guy Standing's concept of the "precariat." Unemployment, short-term contracts and a lack of regulations protecting employees against changes in job positions are underlined as instances of labor insecurity, which is an important aspect of precarization. Another important characteristic, beyond the limits of financial income, is a distinctive structure of social income translated as a vulnerable position resulting from a lack of community support, state, enterprise or private benefits (Standing 2011, pp. 7–13). Accordingly, how social inequalities were reflected in the lives of people who then participated in the protests is discussed mainly by referring to their common experiences of unemployment, poor working conditions, and an inability to satisfy basic needs in various local contexts under neoliberal conditions:

> As in Europe, although in more alarming levels, the Middle East and North Africa (MENA) societies are characterized by a sharp polarization of incomes, the exclusion of a rapidly growing class of educated yet unemployed youth, raising prices for basic commodities, the retrenchment of government services, and, in turn, a growing *precariat* (Tejerina et al. 2013, p. 380, emphasis in original).

Precarity is obviously not a term that merely refers to the living conditions of young people in terms of their employment status in the current socio-economic conditions, but also applies to various groups of society in different stages of life. However, in the case of the recent global uprisings, the concept of precarity has definitely been effective in terms of revealing the common experiences of young participants, particularly with regards to unemployment and poor working conditions, which led them to take to the streets.

The existing literature mostly agrees that young people made up a considerable share of the protesters. Craig Jeffrey, writing on youth in the protests around the world in 2011, states that these movements indicate the crucial need for a globally comparative analysis of the hardships faced by young people such as poor education, inability to remain in formal education due to other pressures, and the widespread phenomenon of youth unemployment, as well as police harassment, corruption, poor housing options, and intergenerational tensions (Jeffrey 2011, p. 146). Young people, while trying to map out a route for themselves under the conditions of neoliberal market capitalism, experience the widening of social inequalities by the absence of income security, the unavailability of decent employment opportunities relevant to their educational background, or their inability to afford basic needs such as accommodation by themselves. As a result, new groups of young people protested against the current austerity policies featuring spending cuts and welfare retrenchment in many European countries (Grasso and Giugni 2016, p. 30). The visibility of young people as the leading figures in protests around the world also strengthened the view that they were the nucleus of this new wave of protests; moreover, due to their active roles in several stages of the protest movements, they have been described as "alchemists of the revolution"[18] (Jeffrey 2011, p. 147). This emphasis on young people's participation in the protests also brought about numerous references to age and generation in terminology describing these mobilizations, such as: the Indignant Generation, Generation Occupy, Generation 700 €, the Desperate Generation, Youth Without a Future etc. (Sukarieh and Tannock 2015, p. 102).

Describing these protests as mere "youth rebellions" and overemphasizing the role of young participants undoubtedly has an inhibiting effect in terms of developing a deep understanding of these social movements by acknowledging the diversity of the protesters' backgrounds and the variety of their experiences shaped by the complex global/local dynamics underlying these mobilizations. Mayssoun Sukarieh and Stuart Tannock state that framing the recent global uprisings as youth-led rebellions is a misinterpretation that "obscures the central role played by a wide range of adults and adult-led organizations, protesting over issues that concern not just young people but people of all ages in society" (Sukarieh and Tannock 2015, p. 107). In a similar vein, highlighting the heterogeneity of the participants' profiles, including organized and unorganized as well as professional and "amateur" activists, Tejerina et al. attentively describe young people's role in the 2011–2012 cycle of contention as "catalysts" without attributing to them the role of being the leaders of the protests:

> These mobilizations were initially called for by core constituencies of young, often college-educated, typically under-/unemployed citizens who articulated widespread grievances and rapidly attracted many other groups, classes and cohorts. It must be noted, however, that younger generations acted as catalysts, igniting but not really 'leading' the protests (Tejerina et al. 2013, p. 384).

These critical perspectives regarding the way young people's participation has been perceived in the recent wave of global mobilization are important in the

[18] Craig Jeffrey uses the phrase "alchemists of the revolution" with reference to its previous use by Marx and Engels in 1978 (Jeffrey 2011, p. 146).

sense that they point to the need to analyze young people's collective action beyond generalizations, assumptions, and short-cut inferences. What is worth noting is that most of these uprisings, including the Gezi protests, emerged in contexts in which young people had previously been criticized and even judged for being apolitical and apathetic—in other words, not interested in what is happening on the social and political level around them (Sika 2012; Sukarieh and Tannock 2015, 102; Tanyaş 2015). Because of young people's remarkable visibility and active participation in the protests, youth then frequently became framed as brave political actors who have the power and dedication to lead a social change, which runs contrary to perceptions of young people as apathetic members of society. Therefore, the way young people's transformative potential and agency have been interpreted in the course of these waves of protests reflects the binarity which, as expected, produces either "celebratory" or "overly dismissive" accounts of youth as revolutionary actors (Sukarieh and Tannock 2015, p. 111). This sort of distorted understanding of youth collective action is far from being limited to the recent protest movements, but rather points to continuities in the social construction of "youth as a revolutionary actor" that asserted itself in the exaggeration of the power and autonomy of youth and the romanticization of youth movements (Sukarieh and Tannock 2015, pp. 111–112). It is crucial to be aware of these trends in studying youth and political action if we are to develop a real and nuanced understanding of young people's participation in collective action. Questioning the binary oppositions along which young people' role in social movements have long been interpreted is not meant "to dismiss or write off the agency or effectiveness of the young as rebellious or evolutionary social actors"; rather, it is intended:

> To insist that youth social activism is put into its broader social and economic context, and to claim that the forms and substance of youth rebellion stem not from some unchanging, universal, innate set of characteristics about youth developmental nature, generational relationships or demographic balance, but are themselves shaped and reshaped according to culture, society, politics and economy of the day (Sukarieh and Tannock 2015, p. 112).

The two common aspects of these recent waves of global uprisings, which are frequently referred to in relation to their young populations, were the role of social media and the occupation of public spaces. Online space, and particularly the extensive usage of social media, has been underlined as one of the main characteristics of the recent uprisings. New communication technologies were undeniably effective in spreading the protests and setting up alternative information and media networks. Moreover, since young people are thought to be highly proficient at using new technologies and developing ways to express themselves in these online spaces, and since they constitute the majority of active users of social media, the central role of online tools and spaces has been seen as stressing the "youth" aspects of these social movements.

In order to determine the shape and form of the new wave of global protests, some scholars regard the role of online spaces as central. Manuel Castells frames communication networks as decisive sources of power-making in a networked society, and highlights the rise of "mass self-communication" with the use of the internet and wireless networks as digital communication platforms as one of the main aspects of the transformation of communication environments. Mass self-communication opens up a space that is difficult for the state and the corporations to control; in other words, it provides people with "the technological platform for the construction of the autonomy of the social actor, be it individual or collective, vis-à-vis the institutions of society" (Castells 2012, pp. 6–7). However, this perspective of ascribing a central role to new online spaces has been criticized for overstressing the influence of these technologies and ignoring their possible drawbacks such as forms of digital exclusion. It is underlined that disparities in access to digital platforms, engagement gaps, and divisions between the consumers and producers of online content, blending racial, class, gender and various other forms of inequalities, all shape online communication (Davies et al. 2016, 14). Therefore, online spaces are not free from inequalities, but rather formed through multifaceted experiences of exclusion. Moreover, "the techno-deterministic" approach is also argued to disregard the role and importance of agency. Without questioning who the protesters are, how they express themselves offline and online in relation to their social positioning, and how they exert power and resistance (Davies et al. 2016, p. 14), framing social media as a central force hinders the possibility of developing a thorough analysis of the movement. Moreover, it should be noted that what gave birth to these movements was not solely the intensive use of social media, but actually the physical encounters with others (Tejerina et al. 2013, p. 383) that were prompted and enabled by online technologies.

The occupation of public spaces as a form of protest has also been discussed as one of the distinctive characteristics of the new wave of mobilizations since it has been frequently witnessed in the recent protests around the world. Tejerina et al. underline that even if radical actions had involved occupation of places of work and study in the past, what was striking in the recent mobilizations was that reclaiming the public space emerged as the most common form of action. Thus the extensive use of this tactic in the Middle East, Europe, and the United States inspired the researchers to frame this new wave of global mobilizations as "occupy social movements" (Tejerina et al. 2013, p. 382). This occupation-based form of protesting undoubtedly reflects a reaction against neoliberal urban policies and represents an action for taking back the spaces which originally belonged to public. Moreover, the occupied places have mostly functioned as the spaces where participants with different viewpoints came together, interacted, communicated and also practiced alternative forms of living. In that sense, the occupied spaces turned into "public spheres" created by the protesters, by which they also developed a collective spirit:

> Across the world, squares and plazas have become *public spheres* where people could not only share alternatives, if not counter-hegemonic discourse, information, viewpoints, and ideas, but also where they could develop a sense of community and incubate novel forms of collective projects and identities. (Tejerina et al. 2013, p. 382)

Taking part in the lively environment of these public spheres, mostly in the form of camps or public meetings, and experiencing participatory democracy, as expected, influenced the ways in which protesters related to social and political issues and particularly how they framed transformation. Encouraging social awareness and political consciousness, these spaces enabled a stimulation of discussions driven by the notion of "common good," which the "apolitical young people" who were then the current participants of the protests were thought to have lacked previously. Tejerina et al. state how the recent wave of protests manifestly reshaped the perceptions of the participants regarding social change, democracy and participation:

> The 2011-2012 cycle of protest gave democracy a new meaning, turning it into a horizontal, deliberative, transparent, and participatory dialogue between 'common persons'—not only activists, or militants—concerned with the 'common good'. In doing this, it demonstrated that another way of engaging with the *public sphere* was possible, and it helped initiate or re-engage vast numbers of citizens previously disenchanted with, or disbelievers of politics as a key mechanism for the transformation of social reality. (Tejerina et al. 2013, p. 383, emphasis in original)

The central role of occupation of public spaces also warrants attention, as it was in these places where protesters laid the foundations of their nascent collective spirit, that kept them active and attached to these social movements. Along with shared grievances upon which the collective action arose, being together in a space with boundaries and reclaiming and protecting this space, in a way strengthened the participants' sense of collectivity. The occupied spaces, and the barricades as the walls of these spaces, defined the "in and out" and "them and us," and therefore provided the physical conditions for the feelings of togetherness regardless of the ideologies and organizational priorities, as Manuel Castells suggests:

> by joining an occupied site and defying the bureaucratic norms of the use of space, other citizens could be part of the movement without adhering to any ideology or organization, just by being there for their own reasons (Castells 2012, p. 10).

As it was discussed in the initial section of this chapter, emotions have long been ignored in the study of social movements. However, thanks to the contributions of the cultural approach, collective action scholars have acknowledged the importance of emotions as co-constituents of cultural practices together with thoughts and morals. The role of emotions in getting people out into the streets to act in concert with others has also been recognized in the context of the recent wave of mobilizations. Emotions such as anger, humiliation, fear, and anxiety are said to be the consequences of social inequality, unemployment, insecurity, and oppression as all of these are commonly experienced under neoliberal conditions, and are therefore highlighted as factors leading people to take part in the uprisings (Castells 2012; Benski and Langman 2013). Castells' emphasis on the individual as the building block of collective action brings with it a focus on emotions. Studying individuals as the participants in social movements inevitably involves an analysis of how emotions shape their ways of relating to and taking part in collective movements, since "at the individual level, social movements are emotional movements" (Castells 2012, p. 13). Writing on the recent wave of protests, Castells frames movements in Tunisia, Iceland, Egypt, Spain and the US, as "networks of outrage and hope" (Castells 2012). In doing so, he accentuates how emotions, having been transformed from negative to positive and communicated between individuals through networks, have driven and shaped the content and form of this global wave of mobilizations. Negative emotions translate into anger and prepare individuals to take action as soon as they overcome their fear of the authorities, and particularly of the potential violent treatment that they may face. Castells gives an up-close description of the way emotions push individuals in social movements and provide the ground for the collective spirit.

> if many individuals feel humiliated, exploited, ignored or misrepresented, they are ready to transform their anger into action, as soon as they overcome their fear. And they overcome their fear by the extreme expression of anger, in the form of outrage, when learning of an unbearable event suffered by someone with whom they identify. This identification is better achieved by sharing feelings in some form of togetherness created in the process of communication (Castells 2012, pp. 14–15).

Phil Mizen, on the basis of a qualitative study of the participants in a local Occupy movement in the UK, underlines the importance of "emotional reasoning" in the process of young people's participation in the protests. Critically engaging with perspectives that assume emotions to be affiliated with irrational and unreasonable behavior, Mizen studies how emotions influence and shape young people's decisions and argues that "emotions do play a clear and decisive role in young people's evaluations of the world and their understandings of their relations to the things that they care most forcefully about" (Mizen 2015, p. 180). Benski and Langman also take a closer look at the combination of cognition and emotion in the context of mobilizations of 2011. They underline the process of emotional liberation, along with cognitive liberation, and scrutinize how these processes function in the subjectification of the individuals. Focusing on how emotions connect individuals to society/authorities, the writers argue that an emotional constellation made up of a particular combination of non-congruent emotions, such as anger, and indignation towards the authorities coupled with hope for social change, stirred up people for mobilization while liberating them emotionally from the ongoing forms of grievance (Benski and Langman 2013, p. 536).

The recent wave of global mobilizations has mostly been analyzed as uprisings of the young precariat who face the harsh impacts of growing social inequality in the context of global neoliberal rule and authoritarian regimes in different societies. Furthermore, young people have frequently been seen as the central figures of these protests taking place in several different national contexts, as members of a generation for whom unemployment, poor working conditions, insecure jobs, unaffordable housing, and weak welfare provisions became the norm, rather than an exception. Even if young people's participation has been visible in most of these mobilizations, which makes it reasonable to analyze their reasons and motives for taking action, it is also important to be aware of the possible consequences of overemphasizing their role and characterizing them as the "revolutionary youth" of this era, as if being oppositional, organizing resistance, and taking action were innate characteristics of youth. The occupied public spaces, as occupation is frequently practiced in many of these mobilizations, have been places in which the protesters come into contact with others in the movement, engage in discursive practices and experiment with alternative forms of living as well as doing politics. In that sense, these new public spaces that emerged in the occupied zones are worth analyzing as sites of transformation and collective action. Emotions, helping maintain the protesters' involvement, have also been included in the analysis. In a way, studies of the emotional background of recent instances of collective action in the global wave of mobilizations have inspired this research's special emphasis of young individuals' reflections on their emotions and how they affected the Gezi protests.

### 3.2.3 Youth in the Gezi Protests

The Gezi protests started in the Gezi Park near Taksim Square in Istanbul in late May 2013 as a protest against the destruction of the greenery in the park to make way for the reconstruction of a historical Ottoman military barracks, with a shopping mall and a mosque in it. In a matter of a few days, the protests spread all over the country. The focus of the uprising broadened, targeting not only the government's urban regeneration policies but also its authoritarian stance and neoliberal policies in general that had affected the lives of individuals in areas such as employment, housing, education, use of public spaces, freedom of expression and organization, etc. From the very start, the Gezi protests were deemed part of the recent wave of global uprisings happening from 2009 on (Davies et al. 2016; della Porta 2015, 2017; Tuğal 2013; Vatikiotis and Yörük 2016). The combination of social actors from diverse ideological backgrounds, the effective use of social networking sites, the occupation of a public space, the development of discursive practices in the occupied zones, and the emerging alternative visions of democracy have been underlined as the characteristics that the Gezi protests shared with mobilizations in other countries. Moreover, a kind of snowball effect was evident in these global mobilizations, especially visible in the reference points of their political contention (symbols, discourses, and practices), as in the case of the Gezi protests with references to Tahrir, Greece, and Brazil (Vatikiotis and Yörük 2016, pp. 2–3). Defining the 2009–2013 resistance wave as revolt against commodification in its multiple appearances, Cihan Tuğal argues that Turkey differs from other mobilizations in the sense that the focus of the Gezi Park Protests was the commodification of nature and other shared spaces, while it was the commodification of money or the commodification of labor that was the crux of the matter in other instances (Tuğal 2013, p. 148). Besides, the Gezi protests have been interpreted within a trend, particularly noticeable in mobilizations taking place on the periphery of the EU and in other developing countries, and articulated as "not so much demands for the radical upheaval of democracy, but rather calls for more just, responsible, and transparent administrations, and the moderation of illiberal, exclusive, and authoritarian policies" (Davies et al. 2016, p. 8 citing Kuymulu 2013). It is also crucially important not to overlook the variety of the issues raised by the protesters in the Gezi protests that were mostly articulations of the interaction between global and local processes and the intersection of collective and individual grievances. Thus the agenda of the protests included a range of issues such as "the precariousness of social life, the refusal to be subordinate to authoritarianism, the will to control the course of one's existence and free it from the influence of the dominant powers, besides the questions regarding the control of the economy, urban development, and the prospects of redefining democracy" (Farro and Demirhisar 2014, p. 181).

The local background to the Gezi protests, especially in terms of the previous mobilizations and the growing visibility of new social movements in Turkey in the period shortly before the protests, also warrants attention in order to understand the global-local contextual dynamics in relation to which the Gezi protests emerged. Based on the protest event data including the period between 2011–2013 formed by della Porta and Atak, people in Turkey were not that quiet until the outburst at Gezi. Protests had already been taking place, focusing mostly on social and economic issues, but also on civil rights, the Kurdish question, labor and environmental problems, and voicing nationalist sentiments or Islamic resentment against suppressive regimes in the Middle East—mostly after the military coup in Egypt or the war in Syria (della Porta and Atak 2017, p. 41). Accordingly, della Porta and Atak suggests that:

> The Gezi Park mobilizations built on a relatively diverse and vibrant protest environment in the country. In our view, this provides a useful indication of the fact that Gezi brought together miscellaneous groups with convergent and divergent stances. The usual suspects of contentious politics in Turkey brought their own claims, repertories, and resources, enriching the collective agency of the Gezi Park protests (della Porta and Atak 2017, p. 42).

Regarding the question of who the participants of the protests actually were, most of the initial interpretations described the protesters as the middle class. Tuğal suggests that "the Gezi Resistance appears to be an occasionally multi-class, but predominantly middle-class movement," in which "generously paid professionals" of the labor market constituted the core group of participants (Tuğal 2013, p. 156). The roots of grievance for this class that caused them to take to the streets were not necessarily economic exploitation or impoverishment, but the impoverishment of social life. Lacking a feeling of fulfillment in their lives, they sought it in the non-commodified space, i.e. the barricades, the public park, and shared meals that the protests provided, and had the chance to experience a way of life based on solidarity, and it was this that kept them in the park (Tuğal 2013, p. 157). Similarly, Çağlar Keyder underlines the central role of the middle-class youth in the protests and described this group as.

> a new middle class in formation, whose members work in relatively modern workplaces, with leisure time and consumption habits much like their global counterparts'. But they also look for new guarantees for their way of life, for their environment, for their right to the city; and they resent violations of their personal and social space. (Keyder 2013)

However, scholars have also pointed out the limitations of class-focused perspectives. Critically engaging with the approaches that view the Gezi protests as solely a middle-class movement, Efe Can Gürcan and Efe Peker suggest that the protests brought together various wage-earning class segments who shared "the common destiny of decreased class and life-chances, and increased precariousness, exploitation and proletarization," mainly composed of service sector employees, relatively qualified wage workers and educated youth (Gürcan and Peker 2015, pp. 18–19). Erdem Yörük and Murat Yüksel claim that class cannot explain the dynamics or the profile of the Gezi protests, and they emphasize the protesters' political and cultural orientations rather than their class positions. By doing so, they approach the Gezi protests as a people's movement based on political demands—rather than strictly class-based ones—in which members from all social classes participated (Yörük and Yüksel 2015, p. 164). In a similar vein, della Porta and Atak problematize the handling of the actually diverse motives of the participants as if they were all about class, highlight the issues of lifestyle, values, and orientations of the participants, and point out the anti-authoritarian component of the protests which served as a common ground for the protesters:

> The erosion of social rights and of economic rewards of education as well as the precarious nature of employment might have activated class motives for protest… Yet, it would be far-fetched to generalize such motives to the entire course of the Gezi Park mobilizations. Articulations that are not compellingly related to class—such as those concerned with lifestyle, values, and orientations, or what Bryan S. Turner (1988) referred to as "status politics"—existed side by side with the class roots of the resentment of some, if not all, protesters (della Porta and Atak 2017, p. 40).

These discussions about the social basis of the protests show that studying the profile of the participants of the Gezi protests by primarily focusing on their class positions carries the risk of obscuring the diversity of their political and cultural backgrounds, and accordingly their different demands that eventually merged during the protests. However, this should not mean ignoring the socio-economic roots of the mobilization. The question as to how the participants' socio-economic positions link with their participation in the protests still stands as one of the most important questions for this field of inquiry.

As mentioned, the participants of the protests were mostly assumed to be young people. Furthermore, similar to their global counterparts, the Gezi protests were frequently framed as a "youth movement," "youth uprising," and "young people's resistance" in the public discourse in Turkey.[19] Quantitative data provides some clues regarding the level of youth participation and the profile of the young participants in the protests. According to the face-to-face survey research conducted by KONDA with 4411 participants in the park area over the course of the protests, 16.5% of the protesters were aged 18–20, 30.8% were aged 21–25, while 20.3% were 26–30 (KONDA 2014, p. 10). This data shows, then, that the majority of protesters in the park were young people under the age of 30. However, taking into consideration that the scope of this quantitative research was limited to the protesters in the park area, it is not possible to apply its results to all mobilizations in several parts of Istanbul and around the country that took place as part of the Gezi Park Protests. Moreover, as discussed in the previous section, overemphasizing young people's role in the recent mobilizations brings along a distorted and deficient understanding of the movements, and more importantly hinders an in-depth analysis of the ways young people become involved in collective action by assigning to them the role of revolutionary actors from the very outset and seeing youth agency as limited to this imagined role. Instead, this research refrains from labeling the Gezi protests as a "youth movement" or a "youth-led movement," and therefore does not attribute a central role to young participants in the protests. Rather, acknowledging the significance of young people's participation and their influence in determining the form and nature of the protests—via the language used in the slogans, chants, and writings as well as via modes of relating to each other in the park etc.—this research focuses on the young people in the protests in order to discover their ways of participating in this collective action. To put it more clearly, the reason for choosing youth as the subject matter of this research does not stem from taking the idea that there is a linear and positive relationship between being young, being political, and being revolutionary for granted; on the contrary, it is rooted in the need to scrutinize being young, being part of collective action, and various ways of relating to the politics in relation with an open and critical perspective by focusing on the case of the Gezi protests.

The collective identity formed in the course of the protests, as it is particularly experienced through solidarity-based relationships, collaboration, and dialogue between individuals and groups who had previously been distant from each other

[19] Pieces by two well-known Turkish columnists can be offered as examples of this sort of framing of young people in the Gezi Protests: (Çandar 2013) and (Özkök 2013).

based on the differences of their worldviews or political tendencies, has been one of the mostly often discussed topics regarding the literature on the Gezi protests. The Gezi spirit, as it is often referred to by the protesters themselves, refers mainly to the specific ways of interacting with one another that the protesters had rarely experienced before the protests and were especially significant when manifested in practices based on the spirit of exchange, gift-giving, politeness, nonviolence, and collective work (Örs and Turan 2015). The exceptional experience of encounters and interaction between the protesters that gave rise to a collective identity has been underlined in terms of its transformative impact on the participants, by way of which they escaped the limitations of their "default" identities and started to become something else:

> It is out of this long-lost feeling of solidarity and commonality visited the park, which is related not to what one is but to what one becomes. Countless individuals bear testimony to the feeling that they have never experienced anything comparable to their days in the park and are transformed by what had taken place. This is the logic of becoming: becoming is based on the seeds of transformation planted in an encounter in which each party moves on to something else that he or she could have never formulated before or outside the encounter. (Eken 2014, p. 434)

The two-week occupation of Gezi Park has been interpreted as a unique experience for the participants both as the site of the emergence of the Gezi Spirit and as the realization of political ideals and experimentation with alternative ways of living and doing politics. By engaging in practices centered around commons such as knowledge, care, and medical aid, the protesters in the park also made a public declaration about how the social should be organized, going beyond an expression of a particular social group's discontentment with the government (Karakayalı and Yaka 2014, p. 125). A cross-cultural, cross-religion, cross-ethnic, cross-class, and cross-gender form of solidarity emerged as an outcome of the civil utopias in the Gezi protests (Kaya 2017, p. 2). Moreover, both the occupation of Gezi Park and the further spread of the local park forums have been emphasized as exceptional instances which point to a potential for radical democratic politics to flourish: "Urban citizens were reclaiming the commons for radical democratic formations while in the process of establishing agonistic relations" (İnceoğlu 2014, p. 26).

Both the transformative encounters between the participants and the formation of alternative public spaces are worth analyzing in terms of their impact on the individual level. The protesters mostly framed their participation in the protests as individual participation[20]—even if they were members of political organizations, and "individual dignity," in other words "controlling the course of one's own existence," was one of the main issues (Farro and Demirhisar 2014, p. 180). Therefore, if and how individuals were influenced and changed by their protest experience is an important topic of research of the Gezi protests. It has been suggested that participation in the protests marked an important turning point between the *before* and *after* Taksim in individual lives (Farro and Demirhisar 2014, p. 181, emphasis in original). The newly emerging political subjectivities have been highlighted regarding the transformative effects of the protest experience for the individuals:

> The Gezi uprisings clearly unleashed transformative effects, at least on the individual level, and set the ground for the formation of new political subjectivities. Unusual but revealing encounters with violent state apparatuses, with the other and unknown dissidents on the street as well as experimentations with alternative imaginaries of politics such as neighborhood assemblies empowered people, broke routines, and let the previously unthinkable emerge (della Porta and Atak 2017, p. 55)

The emergent literature on the Gezi protests has already produced stimulating discussions in the fields of collective action and politics. Works focusing on the contextual and individual reasons for participation as well as the social basis of the protests, in addition to studies addressing the transformative impact of the Gezi Spirit and the political potential of the new alternative public spaces in the course of the protests, all have undoubtedly contributed to our understanding of the events at Gezi. However, there is certainly much more to be discovered about the protests regarding their background, components, participants, and their individual and societal effects. What the literature strikingly lacks is qualitative and in-depth inquiry that establishes the articulations of the diversity inherent in the protests and the complexity of the process of collective action by taking a closer look at particular groups/individuals or specific periods/spaces of protest. This kind of qualitative approach has the potential not only to reveal the heterogeneity of the experiences embedded in the protest process but also to discover new ways of analyzing various political and cultural groups in a multifaceted, dynamic, and complex social movement setting. Some studies that are illustrative of this perspective have already made a significant contribution to the literature

[20] According to the survey conducted by KONDA, 78.9% of the participants were not members of any political party, organization or association, foundation, or platform. In addition, 93.6% of the participants declared that they took part in the protests just "as citizens," that is, not as representatives of any group or entity.

on the Gezi protests. On the basis of qualitative fieldwork conducted with the feminist activists who took part in the Gezi protests, Buket Türkmen investigates the impact of protest experience on the lives of feminist participants, with special emphasis on their long-term transformative effects on female participants (Türkmen, 2014). In a similar vein, Ayşe Deniz Ünan's work focuses on the members of the LGBT rights movement in the Gezi protests. Analyzing the impact of the protests on their LGBT participants, Ünan specifically underlines how LGBT individuals began to feel more confident in their identities and became more politically involved in public demonstrations as a result of their protest experience (Ünan, 2015). Employing an ethnographic perspective, Lorenzo D'orsi studies the lived experience of the activists during the occupation of the park, street demonstrations, and park forums in order to understand how such a heterogeneous mass united into such a large movement, which was something not even the participants themselves expected (D'orsi 2015, p. 17). Another study, by Hrag Papazian, centers on an Armenian group of leftist activist young people who were active in the Gezi protests. On the basis of ethnographic research, he discusses the political subjectivities of group members as well as their political activities and activism through the lens of intersectionality (Papazian 2016). Özge Özdüzen studies cinema-centered creative forms of protest, focusing particularly on film screening activities in Gezi Park and other parks during the Gezi protests. Basing her analysis on ethnographic fieldwork conducted with a cinema audience community that was involved in both the struggle for Emek Movie Theater and the Gezi Park mobilization, Özdüzen traces the processes of cinema-audience activism in these protest periods relationally (Özdüzen 2017).

There are also a number of studies focusing on the category of youth in the Gezi protests that take a qualitative approach. Examining youth political participation through the concepts of political psychology, Emre Erdoğan underlines some of the common themes such as "grievance," "anger," "feeling excluded," and "political cynicism" concerning the roots of young people's participation. He discusses the possibility of the emergence of a collective political identity based on these feelings and emphasizes young participants' unwillingness to transform the Gezi movement into a long-term political movement as a significant obstacle regarding the potential for the formation of this collective identity (Erdoğan, 2015). Bahar Tanyaş investigates how young participants in the Gezi protests perceive the discourse of the nonpolitical young generation and position themselves and their generation in relation to this discourse. She argues that even young people tend to distance themselves from the nonpolitical youth discourse in one way or another; they do not ignore or deny it. They are concerned about the possibility that the next generation might be nonpolitical. Tanyaş concludes that the notion of political apathy still has substantial power (Tanyaş 2015). Yılmaz and Gümüş study the Gezi protests by focusing on young participants from new social movements. On the basis of qualitative field research conducted with feminists, LGBT youth, and environmental activists before and after the protests, they argue that young people's interest in and struggle to influence public matters in the Gezi protests were not unrelated to the youth participation that had already been developing in these social movements (Gümüş and Yılmaz, 2015). Another recent article by Begüm Uzun concentrates on student mobilization in the Gezi protests. Founded on a qualitative study conducted with university students who took part in the protests, Uzun compares student activism in the Gezi protests with the previous student movements in Turkey from the 1960s to 90s. She defines the politics of student activists in the Gezi protests as a "politics of lifestyle," and states that while students in the past had adopted clear political objectives, i.e. pushing the government for reforms in the post-secondary education or initiating a socialist regime, students in the Gezi protests "limited their objectives to being heard and respected by the government as equal citizens of the nation" (Uzun 2017, p. 168).

## 3.3 Conclusion

This chapter aims to set the main theoretical background as well as to reflect on the relevant literature regarding the analysis of young people's experiences of the Gezi protests. Therefore, the first part addresses the main trends which have been influential in the field of social movements and clarifies what kind of approach this research takes in terms of studying the Gezi protests as a social movement. While describing the cultural perspective employed in this work, the long-ignored importance of emotions and morals as indispensable parts of cognition has been primarily underlined with regards to individuals' experiences of collective action. Also, the process-based definition of collective identity and the performative understanding of the course of the protests have been stressed as the guiding conceptual discussions for the field data analysis of this work.

The second part of the chapter deals with the question of how youth has been studied in social movements literature. This part starts by underlining some remarkable instances of young people's collective action on the global scale from the 1960s onward. the chapter proceeds with a review of youth in the history of social movements in Turkey, specifically highlighting the need to understand young people's presence in the emergent new social movements in Turkey in the post-1980 period, since young people's collective actions in the Gezi protests have mostly been interpreted as being rooted in their previous engagement in those social movements. This is followed by discussion of the literature on the worldwide wave of uprisings between 2009 and 2013, aimed to place the Gezi protests in the context of social movements globally. This section also covers the way discussions on the contextual roots and social basis of these protests, and particularly how youth has been framed as the central figure of the protests. It discusses the common techniques used in the organization of the protests, such as social media and occupation of public spaces, as well as the pivotal role of emotions in motivating individuals to join the protests. The last section takes a closer look at the growing literature on the Gezi protests, discussing the socio-economic background of the participants who are thought to be mostly young people similar to their global counterparts, and the process of the formation of collective identity and transformation experienced by individuals within the emerging alternative political spaces. Pointing out the limitations of qualitative in-depth research on the topic, the chapter ends by drawing attention to the necessity of taking a closer look at young people's participation in the protests.

This chapter clarifies the conceptual references relevant for the discussion of young people's participation in the protests and contextualizes youth collective action in the Gezi Protest in national and global terms. As discussed above, one of the main problems regarding the way young people have been framed in the current literature on social movements is the binary categorization of youth and their political agency as either apathetic or revolutionary actors. The perception of young people's visibility in the Gezi protests has also been exemplary of this binary tendency, as young people of the post-1980s generation who had previously been judged for being apolitical were suddenly glorified as revolutionary political actors on the streets struggling for the future of the country. Therefore, one of the primary challenges of studying youth in the Gezi protests is to go beyond this reductionist way of understanding young people's collective actions by scrutinizing young individuals' motivations for and forms of participation in the protests as well as their reflections about the period of the protests, the run-up, and the aftermath. In this way, youth is viewed not as a coherent group and solid entity, but as an age category composed of individuals with a wide range of experiences. This perspective enables us to discover the link between the social conditions of being young and young people's collective action, which is one of the primary aims of this research. Coupled with the cultural-performative perspective in studying social movements, a sociological understanding of youth in the protests is employed not only in hope of creating a comprehensive analysis of young people's participation in the protests but also in order to provide valuable insights into different ways of being young in today's Turkey.

## References

Alper, E. (2010). Reconsidering social movements in Turkey: The case of the 1968–71 protest cycle. *New Perspectives on Turkey, 43*, 63–96.

Altınay, A. G. (5. February 2003). Anti-war Activism in Turkey. *Bianet.* Retrieved January 5, 2022, from https://m.bianet.org/english/people/16512-anti-war-activism-in-turkey.

Baykan, B. G., & Lüküslü, D. (2010). Gençlere Göre Çevre: Küresel ama Satırarası Bir Sorun. In C. Boyraz (Ed.), *Gençler Tartışıyor: Siyasete Katılım, Sorunlar ve Çözüm Önerileri* (pp. 257–271). Istanbul: TÜSES.

Benski, T., & Langman, L. (2013). The effects of affects: The place of emotions in the mobilizations of 2011. *Current Sociology, 61*(4), 525–540.

Bora, A., & Günal, A. (2002). *90'larda Türkiye'de Feminizm.* Istanbul: İletişim Yayınları.

Boğaziçi Üniversitesi Kadın Araştırmaları Kulübü [Bogazici University Women's Studies Club] (2006). *BÜ'de Kadın Gündemi 10 [Women's Agenda at BU].* Spring 2006. Boğaziçi University. Retrieved January 5, 2022, from http://www.bukak.boun.edu.tr/?p=1152.

Boğaziçi Üniversitesi Kadın Araştırmaları Kulübü [Bogazici University Women's Studies Club]. (2006). *BÜ'de Kadın Gündemi 11 [Women's Agenda at BU].* Fall 2006. Boğaziçi University.

Calhoun, C. (2001). Putting emotions in their place. In J. Goodwin, J. M. Jasper, & F. Polletta (Eds.), *Passionate politics: Emotions and social movements* (pp. 45–57). Chicago and London: The University of Chicago Press.

Çandar, C. (3. June. 2013). Postmodern Bir Direniş [A Postmodern Resistance]. *Radikal.* Retrieved January 5, 2022, from http://www.radikal.com.tr/yazarlar/cengiz-candar/postmodern-bir-direnis-1136001/.

Castells, M. (2012). *Networks of outrage and hope: Social movements in the internet age.* Cambridge and Malden: Polity Press.

Clarke, J., Hall, S., Jefferson, T., & Roberts, B. (2006). Subcultures, cultures and class. In S. Hall & T. Jefferson (Eds.), *Resistance through rituals: Youth subcultures in post-war Britain* (2nd Edition) (pp. 3–59). London: Routledge.

D'orsi, L. (2015). Crossing boundaries and reinventing futures: An ethnography of practices of dissent in Gezi park. In G. Koç & H. Aksu (Eds.), *Another brick in the Barricade: The Gezi resistance and its aftermath* (pp. 16–33). Bremen: Wiener Verlag.

Darıcı, H. (2013). "Adults see politics as a game": Politics of Kurdish Children in Urban Turkey. *International Journal of Middle Eastern Studies, 45*(4), 775–790.

Davies, T., Ryan, H. E., & Milcíades Peña, A. (2016). Protests, social movements and global democracy since 2011: New perspectives. In T. Davies, H. E. Ryan, & A. Milcíades Peña (Eds.), *Protests, social movements and global democracy since 2011: New perspectives* (pp. 1–30). UK: Emerald Publishing.

della Porta, D., & Diani, M. (2006). The study of social movements: Recurring questions, (Partially) Changing answers. In D. Della Porta, & M. Diani (Eds.), *Social movements: An introduction* (pp. 1–30). Blackwell Publishing.

della Porta, D. (2015). *Social movements in times of austerity: Bringing capitalism back into protest analysis.* Cambridge and Malden: Polity Press.

della Porta, D. (2017). Riding the wave: Protest cascades, and what we can learn from them. In D. della Porta (Ed.), *Global diffusion of protest: Riding the protest wave in the neoliberal crisis* (pp. 9–30). Amsterdam University Press.

della Porta, D., & Atak, K. (2017). The spirit of Gezi: A relational approach to eventful protest and its challenges. In D. della Porta (Ed.), *Global diffusion of protest: Riding the protest wave in the neoliberal crisis* (pp. 31–58). Amsterdam University Press.

Eken, B. (2014). The politics of the Gezi park resistance: Against memory and identity. *The South Atlantic Quarterly, 113*(2), 427–436.

Emiroğlu, T. (2013). 2000'li Yıllarda Yeni Aktivizm Biçimleri: Yeni Toplumsal Hareketlerde Gençlerin Siyasi Angajmanı. In D. Lüküslü & H. Yücel (Eds.), *Gençlik Halleri: 2000'li Yıllar Türkiye'sinde Genç Olmak* (pp. 224–246). Ankara: Efil Yayınevi.

Erdinç, I. (2013). Sendikal Hareket ile Öğrenci Hareketi'nin Karşılaşması: Öğrenci Gençlik Sendikası Genç-Sen ve Sendikalı/Sendikacı Öğrenci. In D. Lüküslü & H. Yücel (Eds.), *Gençlik Halleri: 2000'li Yıllar Türkiye'sinde Genç Olmak* (pp. 292–311). Ankara: Efil Yayınevi.

Erdoğan, E. (2015). Yeni Bir 'Siyah Kuğu'nun Peşinde: Bir Siyasal Davranış Biçimi Olarak Gezi Protestoları. In P. U. Semerci, B. Erozan, & N. İncioğlu (Eds.), *Siyasetin Bilimi: İlter Turan'a Armağan* (pp. 355–398). Istanbul: Bilgi University Press.

Eyerman, R. (2006). Performing opposition or, how social movements move. In J. Alexander, B. Giesen, & J. Mast (Eds.), *Social performance: Symbolic action, cultural pragmatics, and ritual* (pp. 193–217). Cambridge: Cambridge University Press.

Ezgin, Ö. (2013). Alevi Gencligi, Deneyim ve Talepler: Bir Özne Olarak Alevilik. In D. Lüküslü & H. Yücel (Eds.), *Gençlik Halleri: 2000'li Yıllar Türkiye'sinde Genç Olmak* (pp. 176–194). Ankara: Efil Yayınevi.

Farro, A. L., & Demirhisar, D. G. (2014). The Gezi Park movement: A Turkish experience of the twenty-first-century collective movements. *International Review of Sociology, 24*(1), 176–189.

Goodwin, J., & Jasper, J. M. (2004). Introduction. In J. Goodwin & J. M. Jasper (Eds.), *Rethinking social movements: Structure, meaning and emotion* (pp. vii–x). Lanham, Boulder, New York, Toronto, Oxford: Rowman and Littlefield Publishers.

Goodwin, J., Jasper, J. M., & Polletta, F. (2001). Introduction: Why emotions matter. In J. Goodwin, J. M. Jasper, & F. Polletta (Eds.), *Passionate politics: Emotions and social movements* (pp. 1–24). Chicago and London: The University of Chicago Press.

Grasso, M. T., & Giugni, M. (2016). Do issues matter? Anti-austerity protests' composition, values, and action repertoires compared. In T. Davies, H. E. Ryan, & A. Milcíades Peña (Eds.), *Protests, social movements and global democracy Since 2011: New perspectives* (pp. 31–58). UK: Emerald Publishing.

Gümrükçü, S. B. (2010). The rise of a social movement: The emergence of anti-globalization movements in Turkey. *Turkish Studies, 11*(2), 163–180.

Gümüş, P., & Yılmaz, V. (2015). Where did Gezi come from? Exploring the links between youth political activism before and during the Gezi protests. In I. David & K. Toktamış (Eds.), *Everywhere Taksim: Sowing the seeds for a new Turkey at Gezi* (pp. 185–200). Amsterdam University Press.

Gürcan, E. C., & Peker, E. (2015). A class analytic approach to the Gezi Park events: Challenging the 'middle class' myth. *Capital & Class, 39*(2), 321–343.

Herrera, L., with Sakr, R. (2014). *Wired citizenship: Youth learning and activism in the middle east.* New York and London: Routledge.

İnceoğlu, İ. (2014). The Gezi resistance and its aftermath: A radical democratic opportunity? *Soundings, 57*, 23–34.

Jasper, J. M. (2010). Cultural approaches in the sociology of social movements. In B. Klandermans & C. Roggeband (Eds.), *Handbook of social movements across disciplines* (pp. 59–110). NY: Springer New York.

Jasper, J. M. (2011). Emotions and social movements: Twenty years of theory and research. *The Annual Review of Sociology, 37*, 285–303.

Jasper, J. M. (2014). Introduction: Doing protest. In *Protest: A cultural introduction to social movements* (pp. 1–14). Cambridge and Malden: Polity Press.

Jeffery, C. (2011). Geographies of children and youth III: Alchemists of the revolution? *Progress in Human Geography, 37*(1), 145–152.

Johnston, H., & Klandermans, B. (1995). The cultural analysis of social movements. In H. Johnston & B. Klandermans (Eds.), *Social movements and culture* (pp. 3–24). Minneapolis: University of Minnesota Press.

Juris, J. S., & Pleyers, G. H. (2009). Alter-activism: Emerging cultures of participation among young global justice activists. *Journal of Youth Studies, 12*(1), 57–75.

Kabacalı, A. (2007). *Türkiye'de Gençlik Hareketleri.* Istanbul: Gürer Yayınları.

Karakayalı, S., & Yaka, Ö. (2014). The spirit of Gezi: The recomposition of political subjectivities in Turkey. *New Formations, 83*, 117–138.

Kaya, A. (2017). Right to public space: Social movements and active citizenship in Turkey. *Research and Policy in Turkey, 2*(1), 1–9.

Keyder, C. (2013). The law of the father, Retrieved January 5, 2022, from https://www.lrb.co.uk/blog/2013/june/law-of-the-father.

KONDA. (5. June. 2014). *Gezi report: Public perception of the "Gezi Protests", Who were the people at Gezi Park?*. Retrieved January 5, 2022, from https://konda.com.tr/wp-content/uploads/2017/03/KONDA_Gezi_Report.pdf.

Lüküslü, D. (2005). Constructors and constructed: Youth as a political actor in modernising Turkey. In J. Forbrig (Ed.), *Revisiting youth political participation* (pp. 29–36). Strasbourg: Council of Europe Publishing.

Lüküslü, D. (2014). Cyberspace in Turkey: A "youthful" space for expressing powerful discontent and suffering. In L. Herrera with R. Sakr (Eds.), *Wired citizenship: Youth learning and activism in the Middle East* (pp. 76–88). New York: Routledge.

Lüküslü, D. (2015). *Türkiye'nin 68'i Bir Kuşağın Sosyolojik Analizi.* Ankara: Dipnot Yayınları.

Mardin, S. (1978). Youth and violence in Turkey. *European Journal of Sociology, 19*(2), 229–254.

Melucci, A. (1995). The process of collective identity. In H. Johnston & B. Klandermans (Eds.), *Social movements and culture* (pp. 41–63). Minneapolis: University of Minnesota Press.

Mizen, P. (2015). The madness that is the world: Young activists' emotional reasoning and their participation in a local occupy movement. *The Sociological Review, 63*(2), 167–182.

Neyzi, L. (2001). Object or subject? The paradox of 'Youth' in Turkey. *International Journal of Middle East Studies, 33*(3), 411–432.

Neyzi, L., & Darıcı, H. (2015). Generation in debt: Family, politics and youth subjectivities in Diyarbakir. *New Perspectives on Turkey, 52*, 55–75.

Önen, Y. (2015). The connective power of saying "No To War": The influence of "No To War in Iraq coalition. In *Political Culture of Turkey*. Istanbul Bilgi University, Political Sciences Doctoral Programme, Unpublished Doctoral Thesis.

Örs, İR., & Turan, Ö. (2015). The manner of contention: Pluralism at Gezi. *Philosophy and Social Criticism, 41*(4–5), 453–463.

Özdüzen, Ö. (2017). Cinema-going during the Gezi protests: Claiming the right to the Emek movie theatre and Gezi park. *Social & Cultural Geography, 19*(8), 1028–1052.

Özkök, E. (6. June. 2013). "Beddua yerine hayır duası" [A benediction instead of a curse], *Hürriyet,* Retrieved January 5, 2022, from http://www.hurriyet.com.tr/beddua-yerine-hayir-duasi-23443849.

Özmen, F. A. (2011). Politik Bir Gençlik Kuşağı: Post 80 Alevi Gençliği. [A Political Generation Youth: Post 80 Alevi Youth]. *Sosyal ve Beşeri Bilimler Dergisi, 3*(1), 11–22.

Roberts, K. (2015). Youth mobilizations and political generations: Young activists in political change movements during and since the twentieth century. *Journal of Youth Studies, 18*(8), 950–966.

Sika, N. (2012). Youth political engagement in Egypt: From abstention to political engagement. *British Journal of Middle Eastern Studies, 39*(2), 181–199.

Simsek, S. (2004). New social movements in Turkey since 1980. *Turkish Studies, 5*(2), 111–139.

Standing, G. (2011). *The precariat: The new dangerous class.* London: Bloomsbury Academic.

Sukarieh, M., & Tannock, S. (2015). *Youth rising? The politics of youth in the global economy.* New York and London: Routledge.

Tanyaş, B. (2015). Gençler ve Politik Katılım: Gezi Parkı Eylemleri'nde 'Apolitik' Nesil. *Eleştirel Psikoloji Bülteni, 6*, 25–50.

Tejerina, B., Perugorría, I., Benski, T., & Langman, L. (2013). From indignation to occupation: A new wave of global mobilization. *Current Sociology, 61*(4), 377–392.

Telli Aydemir, A. (2011). *Katılımın e-Hali: Gençlerin Sanal Alemi.* Retrieved May 15, 2018 from, https://ekitap.alternatifbilisim.org/katilimin-e-hali.html.

Tuğal, C. (2013). 'Resistance everywhere': The Gezi revolt in the global perspective. *New Perspectives on Turkey, 49*, 147–162.

Türkmen, B. (2014). Gezi Direnişi ve Kadın Özneler. *Kültür ve İletişim, 34*, 11–35.

Ünan, A. D. (2015). Gezi protests and the LGBT rights movement: A relation in motion. In A. Yalcintas (Ed.), *Creativity and humor in occupy movements: Intellectual disobedience in Turkey and beyond* (pp. 75–94). Palgrave Pivot London.

Uzun, B. (2017). Student mobilization during Turkey's Gezi resistance: From politics of change to the politics of lifestyle. In R. Brooks (Ed.), *Student politics and protest: International perspectives* (pp. 157–171). London and New York: Routledge.

Vatikiotis, P., & Yörük, Z. F. (2016). Gezi movement and the networked public sphere: A comparative analysis in global context. *Social Media + Society, 2*(3), 1–12.

Yalçın, S., & Yılmaz, V. (2014). Gezi Protestolarından Yerel Seçimlere LGBTİ Hakları Hareketi ve Yerel Siyaset. *Iktisat Dergisi*, 525, 84-94.

Yılmaz, V. (2010). Siyasi Örgütlenmelerde Genç Olmak: Kurtarmaya Giderken Yakalandıklarımız. [Being Young in Political Organizations]. In C. Boyraz (Ed.), *Gençler Tartışıyor: Siyasete Katılım, Sorunlar ve Çözüm Önerileri* (pp. 191–205). Istanbul: TÜSES.

Yılmaz, V., & Göçmen, I. (2016). Denied citizens of Turkey: Experiences of discrimination among LGBT individuals in employment, housing and health care. *Gender, Work & Organization, 23*(5), 470–488.

Yörük, E., & Yüksel, M. (2015). Gezi Eylemlerinin Toplumsal Dinamikleri. *Toplum ve Bilim, 133*, 132–165.

# Youth in the Gezi Protests: Background, the Gezi Spirit, and Aftermath

4

This chapter studies how young people give meaning to their participation in the Gezi Protests. How did they decide to take to the streets and continue the protests? How did they feel during the protests? How do they reflect on their participation? How do they evaluate the overall effects of the protests? Following these questions, this chapter look more closely at how young people lived through the Gezi Protests, painting a picture informed by young protest participants. Young people's accounts form a background for the interpretation of their relationship with politics and their notions of the political, both of which will be discussed in detail in the next chapter.

The chapter begins with examination of how young people experienced the social context and conditions forming the background to the Gezi Protests before investigating the modes of participation among young protesters and their motives for joining the protests. The third section takes a close look at young people's accounts of the communal experience during the two weeks of occupation of the park. The final section focuses on the aftermath of the protests and explores the impact and legacy of the Gezi Protests from the perspective of the young interviewees.

## 4.1 Refreshing the Memory: Before the Gezi Protests

This first section of this chapter focuses on the important events and social processes in the lead-up to the protests, as identified by the study participants. By taking a closer look at this question, it studies the way in which young protesters perceived and experienced the larger socio-political context that eventually caused the Gezi Protests, and scrutinizes the background against which young people



P. Gümüş Mantu, *Redefining the Political: Youth Experiences of Collective Action in Turkey*, Kulturelle Figurationen: Artefakte, Praktiken, Fiktionen,
https://doi.org/10.1007/978-3-658-40565-6_4

frame and give meaning to their reasons and motives for participation in the protests.

### 4.1.1 Feeling Unhappy and Oppressed: The Socio-Political Context, Earlier Movements

The interviewees mostly referred to the period before the Gezi Protests by emphasizing that they felt unhappy, mostly hopeless, and oppressed in their everyday lives due to the political situation and the socio-economic conditions they were living in.[1] In this context, people from different social and political backgrounds reported that they shared similar feelings that then led them to participate in the protests:

> No one can say that the Gezi (Park Protests) happened because of this or that. Everyone joined for a different reason. And most of the issues had been accumulating over the last few years. Women, Kurdish people, Armenians, youth in the universities, journalists, and so on, everyone had had enough of this, felt oppressed, and took to the streets in one way or another. Some call it the struggle for city spaces, some call it the struggle for nature, but actually it's all the same struggle. (25, employed, member of leftist oppositional organization)

As the young protesters painted their picture of the conditions before the Gezi Protests, the events/policies they primarily emphasized seemed to be strongly linked to their personal cultural and political backgrounds. For example, feminists mostly underscored politicians' sexist speeches on motherhood, family and work life, and the bills to ban abortion.[2] Interviewees who were members of nationalist youth organizations emphasized the Ergenekon trials in which many military personnel, journalists and politicians had received prison sentences,[3] and their view that public celebrations of national holidays had become a struggle with the

[1] This study does not include a detailed analysis of the political and socio-economic conditions in which young people find themselves. These conditions are mentioned to the extent that young people referred to them in their narratives regarding the protests. However, more specifically, social conditions of being young in Turkey are discussed in Chap. 5: Studying Youth in the Social Context.

[2] For a detailed discussion of the AKP government's gender policies, see: (Coşar and Yeğenoğlu 2011).

[3] For a related news article, see: "Ergenekon: The Court Case that Changed Turkey": http://www.bbc.com/news/world-europe-23581891 (accessed January 24, 2018).

governing party.[4] Environmental activists focused on the ecological destruction of the city as a result of large-scale construction projects, while activists from oppositional Muslim groups underlined the inconsistency between the values of equality, justice, and sharing in Islamic belief and the neoliberal capitalist policies implemented by a government that claimed to protect Islamic principles.

Many study participants also underlined that the Gezi Protests emerged as a continuation of the protests and acts of resistance in the country over the previous five years, all of which were reactions to the socio-political atmosphere of the day. Some of these movements mentioned throughout the interviews are: TEKEL Resistance (December 2009–March 2010),[5] in which workers at TEKEL, a recently privatized state-owned factory producing alcohol and tobacco, protested against changes in their contracts resulting in insecure conditions and loss of pay; people's protests against a coal-fired power plant in Gerze[6]; the Sulukule Platform which was founded by volunteers to witness and take part in the urban renewal project of the historical Roma neighborhood[7]; the May 1 protests, which were followed by harsh government reprisals (in 2012–2013); the women's protest against the proposals for abortion ban (3 June 2012)[8]; and protests against the destruction and renewal of the historical Emek Movie Theatre (in 2013).[9] A young participant in the Gezi Protests draws our attention to the precursors of the Gezi Protests:

---

[4] For a related news article, see: "Police, protesters clash at Republic Day March in Turkey": http://edition.cnn.com/2012/10/29/world/europe/turkey-holiday-clash/index.html (accessed January 23, 2018).

[5] For more on this, see: Yeldan 2010 and Özuğurlu 2011.

[6] Gerze is a district of Sinop, a city in the West Black Sea Region of Turkey. For a related news article, see: "Resistance against coal-fired plant in Gerze":

https://m.bianet.org/english/environment/134340-resistance-against-coal-fired-plant-in-gerze

[7] Sulukule is a neighborhood within the old historic part of Istanbul, which has long been a home to the Roma community. For a related news article, see: "Demolitions Continue in the Gypsy Neighborhood of Istanbul": https://bianet.org/english/minorities/109364-demolitions-continue-in-the-gypsy-neighborhood-of-istanbul (accessed January 24, 2018), "Roma Call for Support Against Destruction of Neighborhood" https://bianet.org/biamag/minorities/103958-roma-call-for-support-against-destruction-of-neighbourhood (accessed January 24, 2018).

[8] For a related news article, see: "Turkish women rally against plans to restrict access to abortion" http://edition.cnn.com/2012/06/17/world/europe/turkey-abortion-rally/index.html (accessed January 23, 2018).

[9] For a related news article, see: "One more protest held for Emek Movie Theatre Project": http://www.hurriyetdailynews.com/one-more-protest-held-for-emek-movie-theater-project-18576 (accessed January 23, 2018).

> The history of resistance did not start here (in Gezi Park), nor did the LGBT resistance movement start here. There was a lot of previous resistance that was successful, and there is a lot of resistance going on now. People in Gerze[10] resisted much better than us, with their anchors and shovels. Resistance is continuing in many places in Istanbul. The Sulukule Platform was a very good example of resistance, even if it failed in the end. (23, student, LGBT activist)

Researchers who have studied the politics of the Gezi Protests have discussed how the previous movements contributed to the emergence of such a sudden and diverse popular protest. Gümüş and Yılmaz, on the basis of qualitative two-stage field research conducted in the run-up and the aftermath of the Gezi Protests with young feminist, LGBT, and environmental activists, argue that young people's interest in politics, which became visible in the Gezi Protests, was strongly related to their participation in those new social movements in recent years (Gümüş and Yılmaz 2015). Erdem Yörük, writing about the historical roots of the Gezi uprising, underlines that the year just before the Gezi Protests was strongly marked by political activism (Yörük 2014). Referring especially to grassroots political activism by Kurds, women, workers, LGBT individuals, students, and Alevis, he states that "considering this history and the fact that Gezi has been characterized by multiplicity, it is safe to argue that all the components of this multitude had already been active before the uprising." (Yörük 2014, p. 422).

Young people particularly emphasized examples of the government's repressive attitudes toward earlier protests taking place before the Gezi Protests. The repressive measures against the protestors in these previous oppositional gatherings were considered one of the factors contributing to young people's sense of being oppressed. A young person who became an environmental activist in the course of the Gezi Protest narrates how she was feeling before the Gezi Protests, saying of the events in previous protests:

> I remember that I was so unhappy a few months before the protests, when the oppression increased; the events in the May 1 demonstrations, Emek Movie Theatre[11]... It was like a stick continuously hitting me on the head, and (the protests) felt like an explosion of this, very much like a relief. (23, employed, right-to-city activist)

What is more, the bomb attack in Reyhanlı near the Syrian border (May 11, 2013)[12] was held up as an incident which caused young people to question the

---

[10] See footnote 6.

[11] See footnote 9.

[12] For a related news article, see: "Blasts kill dozens in Turkish town near Syrian border": http://www.bbc.com/news/world-middle-east-22494128 (accessed January 23, 2018).

status quo and become highly critical of the situation in the country. An environmental activist remembered how he and his friends were affected by the news about the Reyhanlı incident and related that following this traumatic event, he started to question the limitations of restricting his political activities to environmental activism and think of new ways to relate to other people outside of the environmentalist circles:

> Before the Gezi Protests, before the May 1 protests, there was the Reyhanlı incident. After the May 1 protests, people had calmed down in a way. I was in the dormitory when the Reyhanlı incident happened. Everyone was feeling really bad, we were sitting and crying in the room. I am from Kırıkhan, my aunts are from Kırıkhan, which is near Reyhanlı, we could not reach any of them, and there was such fear and dread. And I questioned myself several times that May, we were trying to do something, but we could not reach out to others. (22, student, environmental activist)

### 4.1.2 Neoliberal Rule, Urban Regeneration, and Authoritarian Interventions in Everyday Life

Talking about the specifics of the context that paved the way for them to protest on the streets proved to be a reflexive exercise for young people through which they looked back and analyzed the process they went through in the lead-up to the protests. While rethinking and framing the processes before the Gezi Protests, the interviewees mostly referred to neoliberal and authoritarian policies as determinants of the macro socio-political context, which caused young people to feel trapped and afflicted in many areas of life, and in turn led them to protest on the streets. A young activist recounts how living in an environment shaped by neoliberal policies disrupts people's relationship with nature:

> There are neoliberal policies and there are people who are against these policies. Even if they did not feel that much pressure till today, they were under psychological pressure. Everywhere is full of buildings; there is no place for people. Everything is a front. Looking from the outside, there are flowers etc. but it is just a front, in reality it is actually concrete under all of these things. I think this led to an outburst after a while. People are longing for natural life, it is in human nature. (22, student, feminist and Kurdish activist)

Moreover, the interviewees were aware of the fact that neoliberal and authoritarian conditions were global phenomena, and the Gezi Protests echoed the reactions to these conditions on a global scale. An LGBT activist emphasizes that the Gezi

Protests were part of a global wave of protests, all of which emerged as almost "natural consequences" of similar policies of governments:

> The state of affairs in many parts of the world is much the same. People are on the streets, because the policies of the governments and states include similar discourses. People feel the need to take to the streets as a result of the austerity policies and other similar things even in the states that claim to be welfare states. We were already leaning toward going out on the streets, we would have done so one way or another, and Gezi light the fuse. (23, student, LGBT activist)

As discussed in detail in the previous chapter, the socio-political context of the Gezi Protests and their links with protests in other parts of the world have been discussed by researchers who analyze the protests from different perspectives. It is worth recalling Cihan Tuğal's perspective on the Gezi Protests in the context of the global wave of uprisings between 2009 and 2014 with regard to the interviewees' perceptions of the global roots of Gezi. Comparing the protests in Turkey with the protests in Greece, Spain, the U.S., Egypt, Tunisia, and Brazil, Tuğal examines these cases within the context of the hegemonic global capitalist order—in other words, neoliberal capitalism and American domination—and describes the protests as anti-commodification and anti-authoritarian:

> As the hegemons turn from consent to coercion, anti-authoritarian and anti-war protests spread, and are likely to spread further. Depending on the national context and the timing, anti-commodification protests have been (again contingently) articulated to anti-authoritarian protests. (Tuğal 2013, p. 155)

Situating the national/local reasons of the Gezi Protests within the global anti-commodification and anti-authoritarian movements, Tuğal offers an insightful understanding of how the protests against the destruction of a public park and its commodification and conversion into a shopping mall turned into an anti-authoritarian uprising that harshly criticized the government's authoritarian policies in several areas such as urban planning, social policies, women's rights, and everyday life in general.

Within the wide range of neoliberal and authoritarian policies, study participants specifically highlighted urban regeneration projects underway in many areas of Istanbul as one of the most important issues affecting the socio-political atmosphere in the city as well as their own personal experiences in the urban sphere. A young person describes her feelings about the never-ending construction process in Istanbul:

> I am an Istanbul lover and I want to live here. I have been so unhappy with this process of urbanization... I live on the Asian side, which is full of narrow streets, we cannot even walk in those streets now, since it is all trucks and other stuff, I am very unhappy because of this. I am really upset about this "Istanbul is under construction" campaign. (25, unemployed, no political affiliation)

This account illustrates that participants' grievances and demands related to urban planning were far from limited to Gezi Park itself. Rather, young people were aware of the urban restructuring going on in many parts of the city, and were experiencing the effect of this regeneration of the city in their everyday lives. Moreover, they were witnessing the effects of urban regeneration on the lives of the residents in the regenerated neighborhoods, and how people on various occasions protested against and resisted these acts of urban regeneration/gentrification, which, in a way, could be regarded as the predecessor of the Gezi Protests:

> Actually, before Gezi there were many such things [urban regeneration projects] in Istanbul—Sulukule,[13] Ayazma...[14] But those were in more remote places, and therefore did not provoke so much social opposition. Gezi Park was an extreme event. The state planned to take it. It was no different than other places, and they were going to take it one way or another. And it all happened apparently in the eyes of the public... (24, student, environmental activist)

Drawing on Henri Lefebvre's concept of the right to the city, Barış Mehmet Kuymulu reads the Gezi Protests as resistance to an approach to urban planning that prioritizes capital over the interests of the inhabitants. Furthermore, he points to the shift in the agenda of the protests to civil rights and individual and collective freedoms (Kuymulu 2013, p. 276). He also highlights how protesters against the urban policies came to Gezi with a broader set of critiques and demands related to their rights and freedoms in everyday life.

The projects of reconstruction such as the Emek Movie Theatre[15] and the Atatürk Culture Center[16] as buildings of historical and/or cultural significance, both mentioned frequently by the interviewees, are also examples of regeneration

---

[13] See footnote 7.

[14] Ayazma-Tepeüstü is a neighborhood in İstanbul's Küçükçekmece district, one of the sites of large-scale urban regeneration projects.

[15] See footnote 9.

[16] Atatürk Culture Center is a multi-purpose public cultural center building in Taksim Square and one of the symbolic places of the Republican Period of Turkish history as well as republican ideology. Therefore, the rebuilding plan for the center caused many to oppose it, since the renewal plan was interpreted as an ideological move by the current government against the secular republican history of the country. For a news article regarding the destruction of the

projects that led to mass protests and campaigns. These buildings were symbolic places of the republican history of Turkish cinema and arts, and the plan triggered concerns related to country's republican heritage and secularism. Looking more closely at the Gezi Protests through the lens of "culture", Tayfun Atay suggests that the Gezi Park demonstrations are "cultural and rooted in the concerns of the 'secular masses' about the shrinking of the space for their lifestyle as a result of government pressures" (Atay 2013, p. 41).

The government's discourse and interventions in everyday life were also heavily criticized and listed by the interviewees as one of the main triggers of the Gezi Protests. One of the most frequently mentioned examples is the prime minister's statement that male and female students living in the same dormitories did not comply with Turkey's conservative democratic rule.[17] A second example was a new regulation limiting the sale and consumption of alcoholic beverages at nighttime and on the streets.[18] These examples show that young people's unhappiness and sense of being oppressed were strongly connected to the paternalistic and restrictive attitudes of those in power even on issues such as living arrangements (which were considered personal). These obstacles felt by young people even in the micro spaces can be understood as reflections of a new understanding of youth in Turkey, in the words of Demet Lüküslü, "a new myth of [a] pious generation." Lüküslü observes the emergence of this new myth through her analysis of education and youth policies, and governmental discourses from 2011 to 2014, and claims that the Gezi Protests were a socio-political challenge to the "myth of pious youth and all it entails—from the project of Islamization and authoritarian control in general and gender segregation and alcohol restrictions in particular" (Lüküslü 2016, p. 8). Similarly, Ayhan Kaya, in his discussion of how Turkish society and politics have undergone a drastic transformation under the influence of Islam and neoliberalism since 2002, underlining the policies of the ruling party since 2011, argues that the Gezi Protests were partly "social upheaval against the subtle Islamization of Turkish society and politics" (Kaya 2015, p. 64).

The interviewees' accounts related to the period before the protests reveal that they mostly felt unhappy and oppressed by authoritarian neoliberal policies, in

---

center, see: https://www.reuters.com/article/us-turkey-opera/turkeys-erdogan-angers-critics-with-plan-to-replace-culture-center-idUSKBN1D61ZF (Accessed January 23, 2018).

[17] For a related news article, see: "Female-male students living together is against our character": http://www.hurriyetdailynews.com/female-male-students-living-together-is-against-our-character-turkish-pm-57343 (accessed January 23, 2018); for an analysis of this discussion in the field of youth and politics, see: (Uzun, 2014).

[18] For a related news article, see: http://www.bbc.com/news/world-europe-22653173 (accessed January 23, 2018).

addition to Islamization, which had been increasingly influential in the country in the previous years. Moreover, protests that emerged in reaction to this atmosphere before the ones at Gezi Park seemed to have not only inspired but also laid the foundations for the politics of the Gezi Protests characterized by a multiplicity of standpoints and heterogeneity of experiences.

## 4.2 Reasons for and Modes of Participation

Following the analysis of the background to the protests, this section focuses on how young people took to the streets, or to put it another way, how young people were "moved" to protest.[19] How did their unhappiness, sense of being oppressed, and their critical stance toward the socio-political atmosphere they were living in transform into action in the protests? Were there specific events or moments that pushed them to express their complaints publicly and communally? How did these so-called "apolitical" young people decide to take the risk, stand up to the state, and physically take on the police force?

### 4.2.1 Violent Treatment of the Protesters as "Moral Shock"

A group of activists, united under the banner of the "Taksim Solidarity" platform, had been monitoring the regeneration activities planned in Taksim including the Gezi Park area since before the Gezi Protests began.[20] In the last days of May 2013, activists gathered and took turns camping out in the park in order to prevent its destruction. The bulldozers made the first move to cut down the trees and start the reconstruction process on 28 May, but were met with the activists' resistance. Following this, in the early morning of May 30, 2013, the police forcefully removed the tents and used tear gas against the protesters. The news of the police intervention against the peaceful protests spread widely and rapidly, especially via social media. Most of the interviewees stated that what made them decide to take part in the protests was witnessing the police brutality against the initial protests by environmental activists in the park:

[19] I use the verb "move" here with reference to Ron Eyerman's discussion of "how social movements move" informed by performance theory and already discussed in the previous chapter in Sect. 3.1.1. Cultural Approach: Thoughts, Emotions, Morals and Performative Perspective.

[20] "Taksim Solidarity" was founded in 2012 as a platform against the Taksim Project, which includes the urban regeneration of the Gezi Park area and other places.

> What brought me there was the police violence, that is to say the burning of the tents, dragging people away, violence was the main reason for me. (24, student, no political affiliation)

Both quantitative and qualitative research confirms that observing how people were treated by the police was the main factor that prompted young people to support these activists against the police and take part in the protests. According to KONDA's Gezi Report, which is based on face-to-face surveys conducted with 4411 participants in the park over the course of the protests, 49.1% of the participants decided to join the protests after seeing police brutality (KONDA 2014). On the basis of in-depth interviews with young participants in the Gezi Protests, Emre Erdoğan states that there is consensus among young people that the main reason why the Gezi Protests expanded to such a surprising extent was the police violence against the protesters. Furthermore, it was witnessing this violence that stimulated in young people a sense of injustice (Erdoğan 2015, p. 46). The interviewees' accounts analyzed in the present study also demonstrate that watching peaceful protesters facing police violence damaged young people's perceptions of justice. A young environmental activist's account is telling on this point:

> … If they had not used tear gas against that girl in the red dress,[21] I might not have got that angry. I knew that the issue was the environment; these people went there to protect the trees. If you attack these people, then we react. This is all about the oppressor and the oppressed. If the ones who are chased off with stones and sticks had been the police, then people would have gone there to protect them. Because we mobilize when someone is oppressed. As a young person, if there's a fight and someone is being oppressed, I want to step in on the side of that person. (23, student, environmental activist)

As this statement illustrates, the decision to join the protest for this interviewee seemed to be a simple one of taking the side of the oppressed against the oppressor. The interviewee's account highlights that he would have supported the police had they been the ones who were attacked by the protesters. This shows that young people's decision to take to the streets was not a sign of a negative attitude towards the police or the state per se, but was motivated by the unjust treatment of their peers at the hands of those actors. In the eyes of many young people, seeing the initial protesters' being subject to violent intervention by the police despite using peaceful ways of expressing themselves transformed the issue of

[21] The girl in the red dress was one of the symbols of the Gezi Protests. For a related news article, see: "Turkey Clashes, the Woman in the Red Dress": http://www.bbc.com/news/world-europe-22798827 (accessed January 23, 2018).

deciding to stand with the protests into a matter of choosing between "right" or "wrong". Jasper explains how moral vision is intrinsic to the act of protests:

> Protest is pre-eminently about moral vision, for participants make claims about how the world should be, but is not. Often, protest helps the participant realize how the world is and how it should be. (Jasper 1997, p. 135)

In that sense, violent police intervention in the initial days of the protests marked a turning point for the Gezi Protests, since as it caused what Jasper calls "moral shock" (Ibid, p. 137). Witnessing and experiencing police violence had a devastating effect on young people's views on morality and prompted them to join the protests and continue to put up resistance. In other words, being exposed to violent treatment only served to deepen the "moral shock" and feelings of disappointment that had already been developing among young participants in the Gezi Protests. One of the activists describes how she felt when the police mistreated them in the course of the protests:

> The police were giving warnings so harshly, "move on, go home, do this, do that," humiliations, scolding, shouting... That is to say, it feels like we do not live like humans, as if we do not have any value, we are oppressed, subjected to swearing and insults, one feels a bit, I do not know, I am a human, I have a life, I have an existence... (21, student, feminist activist)

The emphasis on morality is also important in terms of explaining how many people from different political backgrounds came together on the basis of common values. Moreover, for young people who intentionally keep themselves at a distance from politics, morals and values seemed to function as a way of relating to the political beyond the constraints of traditional politics. Moving beyond the dualities and limitations of ideologies/discourses of political groups and parties, a simple cause such as saving the trees created opportunity for young people with different perspectives to meet for a common cause at a public park. The statement below is an example of how they framed their participation in the protests beyond the limitations of the definitions of conventional politics, underlining the universal character of the Gezi Protests:

> Because there was an incident there… The police were using excessive force, tear gas even on women and children. These people were there for the environment; they did not have any other goal. These people could have been right wing or left wing but the aim was to save the trees there. This is something universal, people are protesting around the world in order to protect the trees. It is like that in Brazil, in Turkey, everywhere. (23, student, environmental activist)

### 4.2.2 Spontenaous Participation

Previous research has underlined that many participants in the Gezi Park mostly came to the park as independent individuals. According to the survey from KONDA's Gezi Report, 78.9% of respondents were affiliated with neither a political party nor a non-governmental organization such as an association or a platform. Moreover, 93.6% of the respondents declared that they went to Gezi Park as ordinary citizens, not representing any group or formation. Therefore, it can be argued that even if people had affiliations with political groups, they did not take part in the protests as representatives of those groups but as individuals with concerns (KONDA 2014, pp. 16–17). In a similar vein, the interviewees' accounts of what moved them to get out on the streets reveal that many of them did not plan to participate in the protests beforehand but found themselves protesting with others spontaneously. One of the participants narrates her first day like this:

> I was at my friends' house, while we were following the news there from Facebook and Twitter, someone started to play pots and pans at their windows, then we joined in, but we did not have any idea such as going out. Then suddenly, yes we are going, pots and pans in our hands, we went out as a big crowd, much bigger than we had anticipated … I remember this day as the first day. This is the first day I encountered the police and the TOMA.[22] That was the first time I saw how violent all this can be. (23, student, feminist activist)

The interviewees' accounts show that their participation in the protests was impulsive. However, this spontaneity does not mean that there was no background to their participation. Rather, as discussed in the first section of this chapter, young people's grievances and reflections regarding the socio-political conditions in the period approaching the protests are strongly linked to their decisions to express opposition. The interviewees did not frame their reasons for participating in the protests by referring to political definitions, ideologies, and tendencies, they rather emphasized the moral outrage they had felt at the intervention of the police against the initial group of protesters. Therefore, they primarily based their move to protest on moral values and choices. Morality and values were not only effective in the beginning of the protests, but, during the two-week period of camping in the park, also in the way the Gezi Park community attempted to realize the utopia of communal living based on solidarity, sharing, and equality. Moreover, this process itself, including the beginning of the protests and the

---

[22] TOMA is the Turkish abbreviation for "intervention vehicle for social incidents.".

camping period, seemed to open up a space for individuals to reflect on and shape their notions of right and wrong and their idea of what a good world should look like. In that regard, it could be argued that it was this moral ground that kept young people in the protests for a long time, along with the sense of fulfillment that came with struggling for a better life/world/country. As James Jasper states insightfully, morality does not provide "the satisfaction of actually achieving a utopia, but that of striving for one" (Jasper 1997, p. 135).

## 4.3 Camping Out at the Park: Rituals of Solidarity, Emotions, and the Gezi Spirit

This part takes a closer look at young people's experiences in the two-week camping period in Gezi Park, which began following the initial days of protests on the streets. This period has mostly been defined as "Occupy Gezi" in reference to Occupy movements in many other cases (Kaya 2015 and 2017; Bee and Chrona 2017). In addition, many have named this genuine experience of collective living "the Gezi Commune," referring to examples of revolutionary movements such as the Paris Commune.[23] Regardless of the choice of definitions and names, it is beyond doubt that experimenting with communal living with other protesters or groups, including cooperation in practical matters and continuous dialogue between individuals and groups, shaped young people's experiences of the protests.

The components of the collective identity formed in relation to the sense of belonging to the Gezi community will be discussed below, by focusing on how this period was framed by young people. Similarly, the specific features and characteristics of the Gezi spirit that made this collective unique in the eyes of the protesters, will be scrutinized in detail. Emotions and value-based accounts of young people will be highlighted as pointers toward a cultural understanding of how they defined and redefined their notions of politics and the political in relation to their experiences of communal living and struggle at Gezi Park.

[23] Umut Yıldırım criticizes comparisons made by the alternative press between the Paris Commune and the Gezi Commune as tenuous, since "organic links forged between the university students and worker unions during May 1968 did not blossom in the case of Gezi," see: (Yıldırım 2013).

### 4.3.1 Solidarity as the Ground for Collective Identity

What the interviewees often underlined about the days they spent at Gezi Park was the solidarity-based way of living. Solidarity was described primarily in reference to practical matters such as the collective kitchen, the collective library, collective cleaning hours, and the sharing of all the supplies they had such as tobacco, water, or blankets, etc. Furthermore, an understanding of solidarity referring to relations between people at the park was reported to be one of the most important characteristics of the Gezi Park community. Protesters' emphatic efforts to communicate and get to know each other more closely, even if they came from very different socio-political backgrounds, was described as one of the most distinctive features of the Gezi Park experience. A participant recounts how impressive it was for her to witness and experience this as a resident of Gezi Park:

> It was possible to talk about real solidarity in Gezi Park. I am not only talking about food, free tobacco, or Revolution Market.[24] Actually, I had the feeling that people were listening to each other. There were many flags of different parties. The leftist groups, the LGBTs or the Muslims were there, all had their tables close to each other; they were in solidarity, sharing their stuff. I saw all that. (24, student, no political affiliation)

The way they perceived this atmosphere of solidarity and dialogue in the park also had to do with their generational experiences of living a highly individualized way of life, in other words, their substantial remove from collectivity and communal ways of living.[25] The contrast between the "lonely" routine of ordinary life and the "togetherness" of life in the park seems to have had a transformative effect for young people:

> Everyone was in his or her isolated world before the protests, then during the 15 days period in the park people suddenly connected with different people of different views; there were LGBT people, vegans... Everyone was sharing something in solidarity, people became aware that they were not so distant from each other. (34, employed, anti-capitalist and anti-militarist activist)

---

[24] Revolution Market is the name of the tent inside the park area to which food and drink were brought to by people and distributed for free to the protesters on the basis of solidarity.

[25] Individualism is discussed in Sect. 5.1.3. Being Young in Late Modernity: Some Reflections.

As the interviewees repeatedly indicated, the variety of the socio-political backgrounds of the participants was one of the most significant characteristics of the Gezi Park Protests. Especially considering the unexpected coexistence and cooperation between such different groups as leftists, nationalists, Kurdish groups, feminists, LGBT activists, environmental activists, Armenian, Alevi, oppositional Muslim young people, football fan groups, individual participants, and many others, the Gezi Protests presented a novel form of diversity for many. The need to adopt new perspectives—in other words, "spoiling memorization"[26]—in order to understand the politics emerging from this diversity and practiced as being together with differences, has been underlined by Yael Navaro Yashin:

> These protests have not allowed themselves to be placed in any known frameworks of analysis such as secularism versus Islam, modernists versus traditionalists, liberals versus conservatives, the bourgeoisie versus the working classes, left versus right, cosmopolitans versus nationalists, feminists versus sexist men, gay-rights activists versus homophobes. The protesters in the square represented all these inclinations, as well as none. Or rather, the way they expressed their sentiments and desires, spoke their minds and hearts, fitted with no predisposed lenses (Navaro-Yashin 2013).

Therefore, it is hard to explain this unfamiliar multiplicity of different inclinations just as a coalescence of various, either conflicting or unrelated, social and political positions. Rather, the complex process of the formation of collective identity in which these differing tendencies are all involved seems worthy of analysis aimed to gain a deeper understanding of the new politics that have emerged from the Gezi Protests.[27] More specifically, there is an obvious need to understand the Gezi spirit, referring generally to the unique experiences and practices of the protesters in Gezi Park, and the way it was "produced as an interactive and shared definition that is continually negotiated, tested, modified, and reconfirmed by several individuals and groups" (Özkırımlı 2014, p. 4). Within this framework, studying how young people experienced being part of this collective identity bound together by the Gezi spirit emerges as one of the main paths to acquiring a deeper understanding of the ways in which young people related to the political in the case of the Gezi Protests.

---

[26] As described by Yael Navaro Yashin in her article: "Spoiling memorization (*ezberi bozmak*) is a Turkish idiom. It refers to re-doing that which is taken-for-granted. It implies creativity, innovation…".

[27] The constituent parts of the new politics emerging from the Gezi Protests are discussed in detail in Chap. 8.

### 4.3.2 Practicing Utopias

The camp period was crucial for the formation of a collective identity, since it provided a unique opportunity for the participants to relate to each other more closely than they had in the protests on the streets,[28] exchange ideas, and put their ideals into practice. For many, this period spent in the park was full of new and unexpected experiences, and therefore brought about new discoveries about the practical aspects of their political worldviews as well as how these different worldviews related to each other. Moreover, camping in a public park as part of a protest where the protesters were trying to meet everyone's basic needs such as food, shelter, hygiene, and even security on their own, was a physical and material situation that most of these young people had not had a taste of before. A feminist activist shares how living in one of the central public locations in the city was a completely new experience for her. In addition, she underlines how for them, as feminist women, being in the Gezi Park resulted in new kinds of encounters in several spaces, especially with men:

> We had been going to the protests as feminists before... We had also attended mixed protests or organized protests ourselves as the women's movement. However, transforming a space was an all new experience in itself. Collaborating with men, the need to remind the others that you are also there included in this division of labor from the infirmary to the security issues or the kitchen, etc. So, yes, for a participant from the feminist movement, it was especially novel in terms of the interaction with the state, the police, and with men etc. (29, employed, feminist and leftist activist)

In her article based on qualitative accounts of feminist participants in the Gezi Protests, Buket Türkmen draws attention to young women's attempts to disseminate feminist perspectives and fight against gendered practices, in other words, their "womanizing the resistance and the city" throughout the protests (Türkmen 2014, pp. 25–27). She argues that, although a "female urban space" emerged from the spirit of solidarity during the protests, in which young women felt safer and experienced more freedom, especially at night, for young women that period was still marked with struggle against patriarchal discourses and practices (Türkmen 2014, p. 25). The above account also demonstrates that for young women the park experience was especially unique with respect to their struggles to deconstruct gendered norms and practices of the social space, as in the case of transforming

[28] Here it is important to note that it is not possible to make a clear distinction between the camping period and the street protests, as is often the case with occupation and protest movements. While some protesters were in their tents at the park during the occupation, many were also protesting on the streets.

the patriarchal understanding of division of labor, for example, questioning the given association of cooking with women, and security issues with men. Even if life in the park was intended to be organized by the protesters themselves on a solidarity-based structure where official policies no longer applied, that did not make it automatically free from the established patriarchal hierarchies and attitudes in everyday life. Rather, gender roles and norms were, not surprisingly, transferred and put into practice in areas that claimed to be based on a different spirit. Therefore, what the park experience provided specifically to young women, as they recall, seems to have been the moments in which they practiced their feminist politics directed towards the transformation of patriarchal norms in an unusual setting characterized by the coexistence of many different political standpoints.

Regarding this emphasis on the practical side of politics, it is feasible to argue that the park functioned as a laboratory in which different notions of the political and different understandings of an ideal world were lived out. Moreover, this laboratory was also a site where a new collective identity was shaped and performed, which necessarily brings about new ways of thinking and doing politics. Solidarity, sharing, or listening to each other can all be interpreted as practice-based milestones of a collective identity in the making. An ecological activist underlines how their ideas and discussions on the abstract level became real at the park:

> All the things that we used to discuss—alternative economy, solidarity economy, city gardens, etc.—all of a sudden became real in Gezi Park. The small world we imagined started to come into existence. There was labor, but money was not valid, and food was free. (24, student, environmental activist)

This quotation describes the park as a space in which the realization of a prototype of a long-imagined world was possible. The great variety of social and political positions that were represented in the protests undoubtedly helped enrich the protesters' conceptions of an ideal world, their utopias, and also diversified them to a great extent in line with their political and moral visions. In the above statement, the environmental activist stresses alternative economy, solidarity-based economy, and city gardens as important features of her ideal world. In a different vein, an activist from an oppositional Muslim group narrates his experience at Gezi Park with reference to the representation of heaven in the Qur'an:

> It was the twenty-day process we experienced in Gezi, as we call it, a rehearsal of heaven. Addressing Adam and Eve, [the Qur'an] says "Live in heaven, no one is hungry there, no one is burned by the sun there, no one is homeless there." That is to say, the word heaven actually refers to a place where everybody's basic needs are met and everything is shared. (33, employed, oppositional Muslim)

The interviewees' accounts provide strong insights into the extent of the impact that the Gezi Park experience had on the participants. The analogy between heaven and the collective life in the park stands out as a way to frame one's involvement in the Gezi Park community. Moreover, life at the park is mostly described as an extraordinary experience far removed from their everyday life routines. While in this account this extraordinary quality is mentioned with reference to the otherworldly life, in the interview cited below, living in Gezi Park is recalled as a dream that inevitably ended and left the participants in the reality of ordinary life:

> There was the culture of sharing. Everything functioned well; there was the health group, the kitchen group, and the cleaning group. We all had our gloves on and picked up the garbage every morning. It was really a camp site, a huge camping zone and there was a wonderful communal life. When I think of it now, this was a very special part of my life as if I lived in a different place, and then I woke up and came back to the real life. That was like a dream, a one- or two-week dream for Turkey or Istanbul. (21, student, LGBT activist)

### 4.3.3 The Gezi Spirit

Even if young people's descriptions of the ideal world and the concepts through which they reflected on their experiences of collective life in the park differ, what they all similarly emphasized were solidarity and sharing as the founding principles of the park commune. In addition to these, young people underscored some other common attitudes and modes of relating they observed during the period at the park that kept them together and afforded them novel and extraordinary experiences. A participant who became an environmental activist after the Gezi Protests emphasized the distinctive way people related to each other through kindness and care. She also pointed out that people in the park were able to set aside differences and focus on meeting each other's practical needs, such as assistance and information, access to food and collecting information on everyone's blood type in case of an emergency.

> When I was little, if someone bumped into me while I was walking, I usually said "pardon" but the one who had bumped into me would go straight ahead. In the park, there were thousands of people but no one bumped into anyone. I was surprised. Or no one asked one another, "Where are you from?" If someone asks about my origin, I say I am an adopted child, my origin is none of your business! No one asked about my origin in the park, people asked my blood type or whether I was hungry. (31, unemployed, environmental activist)

While talking about novel and poignant experiences in the Gezi Protests, a protester recalls an anecdote from the time:

> I had forgotten to close my backpack while I was running out of the park; I think I was trying to find something, water or something else. While rushing in the middle of this turmoil, I heard the noise of the zipper on my backpack, I had left it open, and someone had closed it while I was running. I have so many little memories like this. (24, student, no political affiliation)

These examples, which otherwise could be considered irrelevant fragments of people's memories of the protests, reveal very important features of the collective spirit that formed around the Gezi Protests. Beyond opposition to and demands from the governing bodies, what kept young people together throughout the protests seems to have been the realization of a mode of living that was expressed in the way people related to each other as well as to their environment. Respect for the trees in the park brought with it young people's respect for others. Therefore, the Gezi spirit, which was mainly based on solidarity and dialogue among the participants, was characterized by attributes such as awareness of others' needs, watching out for others while moving together, and peaceful, harmonious, and kind behavior. A right-to-the-city activist, who became a permanent member of a neighborhood initiative after the protests, underlines that these caring and considerate modes of relating to each other that emerged during the Gezi Protests were something young people felt was lacking in their daily lives, and they had been imagining and looking for this "spirit of fraternity" for a long time before they eventually found it in the protests:

> The best part of the Gezi for me was, even if it was too crowded and people were colliding with each other, then everyone was apologizing and kissing each other. People in Turkey are usually rude to each other on the streets; someone bumps into you and then swears and goes away. People are nervous and somehow there is a lack of cultural development. However, a different culture came into existence there (in the park). Everyone was saying sorry, offering tea and tobacco to each other, it was unbelievable, something that I had never seen before, no one had ever seen anything like it.

> We saw the spirit of fraternity that we wanted to see there. (23, student and employed, right-to-city activist and member of a neighborhood solidarity initiative)

The quotation above shows that the process in which the Gezi spirit was shaped as a collective identity strongly relied on offering young people an alternative in the micro-spaces of their individual lives. By extension, this makes it interesting as a means of understanding young people's perceptions of politics. İlay Romain Örs and Ömer Turan argue that an unwritten code of conduct, which they describe as the "manner of contention," realized in the ethos of collective work, spirit of exchange and gift-giving, politeness and non-violence, provided the ground for peaceful cultural and political pluralism in Gezi Park (Örs and Turan 2015).

With respect to the cultural approach that the present study follows in understanding the Gezi Protests and young people as its participants, paying attention to the role of in the formation of the "we" at Gezi is fundamentally important for the analysis of this process. Studying the emotional background enables an in-depth understanding of how individuals developed attachment to a group of people/movement/action that kept them involved. As discussed in the first part of the previous chapter, emotional processes are frequently key to discovering how collective actions emerge, as well as how individuals become involved in social movements. Emotions have a tremendous impact on the way people think and act in general.

An environmental activist narrates an anecdote in which the protesters worked in the solidarity chain for hours at a stretch. What she underlines is that they were feeling so happy in spite of tiredness, hunger, and danger:

> One night, I became aware that I had been in the chain for eight hours, taking from the right giving to the left. The previous day, they [the police] had attacked so harshly, CNN International broadcast it live for hours. It was so bad. Then some other day we started the chain at 7 in the evening till 2:30 in the morning. The only thing that I was doing was taking from one side and giving to the other, but we were so happy. Cigarettes and chocolates were handed out, and these were our only sources of nutrition. We felt like we could stay there like that all summer, we could live there. People brought anything they had, olive oil, carpets, and tents… I believe that we can form this solidarity chain anytime, since we have witnessed, experienced, and formed it once. (31, unemployed, environmental activist)

If we recall young people's accounts in which they reported feelings of unhappiness, oppression, and hopelessness in the period before the protests, we see in this account an obvious contrast in terms of the way young people felt. Participating in the protests, in the lively flow of the collective action, and experiencing the power of solidarity seem to have had a transformative effect on young people's emotions. Therefore, even if they were exhausted, they felt much better than in the routine of their daily lives. Feelings of trust and security were built along with a sense of happiness and excitement. A participant tells how he felt secure as a member of the collective, and started to trust others:

> We usually walked in groups but we still felt insecure, but it was not like that there [in the protests]. Even if I went alone, I was not afraid, because everyone was helping and being so kind to each other, and I learned to trust people much more. That is to say, I have tasted it. I have experience running away with other during a police intervention. You are together with everyone, you know nothing will happen to you. I had never seen that many people together fighting for a cause before, it was such a beautiful experience for me. (24, student, environmental activist)

Another interviewee recounts the contradiction of still feeling safe even in the debris left behind by the clashes between the police and the protesters:

> There were barricades, broken glasses everywhere on the streets as I was walking. There was still the smell of the gas but I felt much more safe, because I thought that there were people around who would protect me, it was an environment I could not have dreamed of before. (22, student, environmental activist)

These accounts in which young people emphasize their emotional experiences of the protests prove the existence of a substantial relationship between the way participants in the protests felt and the formation of collective identity. Expressions of confidence, security, and contentment functioned as the cement of the collective identity. These emotional articulations emerged and were reciprocally strengthened as a result of the sense of belonging and attachment to this collective identity.

Beyond the common points underlined here with regard to young people's experiences of the protests, the camping period, and especially the way young participants related to the collective identity formed around the Gezi spirit, it is important to note that each and every participant's story and involvement is unique, with its own plot, nuances, and inconsistencies. Young people with different life experiences undoubtedly had different experiences of the protests, as well as different ways of reflecting on their protest experiences:

> Actually, everyone's prior life experiences have been crystallized there. There were people who had lived in a village, or made a camp or lived in a villa... For example, there were those who could not imagine that a collective kitchen could be founded. We were learning. It was also my first experience of that kind: How can one live in

> a park in the biggest square of the most crowded city in the country? (27, employed, leftist activist and member of neighborhood solidarity)

A participant wearing a headscarf reports how she went out into the streets in the initial days, but then did not want to continue to come to the protests. Her story is an example of how the protests were not experienced in the same way by all participants. While for many the Gezi spirit refers to a collective identity they felt connected to acted in accordance with, for some, it had a discriminatory face. She decided to withdraw from the protests due to not having a sense of belonging to the protesters, and sometimes even feeling excluded from the protests as a pious Muslim woman. She first recalls the scene she faced while entering the park with men signing songs and drinking, which caused her to feel alienated from the very beginning as a pious Muslim woman who would not feel comfortable in such an environment. In addition, the "special treatment" of, or "the overly kind" attitude toward oppositional Muslim participants, and especially women, by those with more secular tendencies is given as an example of the events that caused her to distance herself from the protests:

> When I went into the park, the first thing I saw was men singing with beers in their hands... I never felt included and comfortable. Then I experienced one of these "congratulations" incidents; "thanks so much to our friend with the headscarf" or "congratulations for you being here" etc. I gradually became alienated from everything. (25, part-time employed, feminist and oppositional Muslim activist)

These expressions of gratitude to the pious Muslim protesters for their participation betray a hierarchical way of thinking in terms of the co-existence of participants with differing views on religion. In this way, while religious Muslim participants were positioned as mere "supporters" of the protests, others were imagined as the "original protesters" in the park. Moreover, such a perception has specific implications for female Muslim participants and therefore, not unexpectedly, has gendered consequences. The statement below depicts how pictures taken of Muslim women wearing their headscarves was in a way an instance of them being portrayed as "exotic" figures in the context of the protests. An oppositional Muslim activist relates the moments in which they started to take pictures of secular protesters in response to being photographed by them as "interesting objects":

> It was really interesting to be a woman with a headscarf at Gezi Park. Even when you were alone, there were people coming and congratulating you or requesting an explanation from you. However, once, we were together as 10–15 people to shoot a video,

> around 10 of us were wearing headscarves. It was really funny, everyone was taking pictures of us, and then I took the camera and started to take photos of those who were taking our pictures. They were coming and congratulating us for being there and supporting them... And then we said that we are also congratulating you for being here and supporting us. It was like that. (23, graduate student, oppositional Muslim activist and leftist)

A telling moment occurred when Muslim activist groups' turned the camera on those who were taking their photos and congratulated them back. They thereby rejected being objectified as an "attraction" in Gezi Park and reminded others that they were not only supporters but among the main actors of these protests. The fact that these sorts of confrontations occurred at Gezi Park illustrate that even the Gezi spirit, characterized by notions of solidarity and empathy between its components, was not completely inclusive in all stages. However, these examples of exclusionary practices did not make it impossible for these groups to act as a collective, mobilized by the Gezi spirit in the developing stages of the protests. For example, during Ramadan,while the camping period continued in the park, public *iftar*s[29] along the İstiklal Street in Taksim were organized by oppositional Muslim groups and the Gezi Park protesters—most of whom might not have been so religious as to fast, but took part in these *iftar*s as an act of solidarity.[30] Unraveling the boundaries between religious and secular life, *iftar* on İstiklal Street has been interpreted as a "new political space" where the boundaries of social space imposed by normative power were transcended by combining and blurring what was supposed to remain separated (D'orsi 2015, p. 24). Even if identity politics functioned for the Muslim groups in terms of fostering movement aims during the protests, their demands for redistribution and recognition were much more prominent (Dursun 2015, 387). That is to say, as oppositional Muslim participants in the protests, they criticized the neoliberal capitalist policies, defended the right of the people to save a green place in the city, and sided with the oppressed. In this way, they paved the way for meeting other groups on common ground. Therefore, it is important to underline that what made the Gezi collective identity distinctive seemed to be very much related to how conflicts and discrepancies between groups with different political backgrounds were handled during the protests. Similar to young pious Muslim participants' experiences, young feminist protesters' struggles with the gendered perspectives and

[29] Iftar is a dinner during which the daily fast is broken in the month of Ramadan.

[30] For a related news article, see: "Hundreds gather for iftar dinner on main İstanbul Street," http://www.hurriyetdailynews.com/hundreds-gather-for-iftar-dinner-on-main-istanbul-street-84227 (accessed April 6, 2018).

actions of the protesters in the park mentioned earlier in this chapter, also exemplified how the collective identity was formed in the Gezi Protests, in spite of fundamental differences between groups, and more importantly, through an active and reflective engagement with those differences. As a result of ongoing critical communication between female and male participants in the protests regarding the patriarchal discourses and practices, not only did feminist women continue to take part in the protests without making political compromises, but the protests eventually became less discriminatory in terms of gender. Likewise, it seems that the objectification of pious Muslim participants turned into a relationship based on dialogue and solidarity. These examples show that, together with empathy, collaboration, and sharing, tensions, conflicts, and confrontations were also important processes which need to be scrutinized in order to understand the formation of a collective identity. Furthermore, in this framework, a processual and dynamic approach to collective identity, as discussed in the previous chapter, is indispensable for grasping how individuals and groups in a collective relate and become attached to each other throughout these complex and multifaceted processes and gradually build a shared understanding and solidarity among themselves.

## 4.4 After Gezi: Transformation, Political Consciousness, and Disappointment

This final section of the chapter centers on the post-protest period and scrutinizes how young people look back on and evaluate their protest experiences. How does having participated in the Gezi Protests affect their lives? How do they analyze the societal effects of the protests? What has remained after the protests? Instead of limiting itself to the categories of success or failure of the protests, this study favors a more complex interpretation of the young participants' accounts of the effects and the results of the Gezi Protests. Accordingly, this section considers a range of themes such as the transformation of relations between individuals as well as within groups, increasing interest in politics and political consciousness, and diffusion of the transformative effects of the protests in several areas of everyday life, as well as feelings of disappointment and trauma remained after the protests. By scrutinizing the impact of the protest experience on several different interviewees, this work studies the marks that involvement in the Gezi Protests has left on young people's lives.

### 4.4.1 Transformation in Relations and a Developing Sense of Community

One of the most important points highlighted by the interviewees regarding the impact of protest participation on their lives is the change in the way they relate to each other. The experience of communal living in the park and getting a taste of the Gezi spirit, seems to have encouraged young people to transfer into their everyday lives this fresh knowledge of collectivity. An LGBT activist describes how their protest experience developed their ability to form more open and emphatic relationships and therefore real dialogues between individuals:

> What happened after Gezi? The main achievement is not the prevention of the destruction of the park to build a shopping mall in its place. The main achievement is that people started to connect with each other, learn others' viewpoints, and change their attitudes towards each other... What was important for me was this change. (25, student, LGBT activist)

The transformation of one's way of relating to others was mentioned by a number of interviewees as one of the most important impacts of the Gezi Protests. Similar to the above account, another interviewee, a white-collar corporate employee without political affiliation, describes the transformation in her attitude towards others triggered by her experience in the Gezi Protests in the statement below. What is important to note is that this participant was also an attendee at the forum meetings initiated in her workplace following the protests, and this forum experience seems to have had a considerable impact on her in terms of developing her ability to connect with others. She emphasizes the value of the interactive critical discussion platform created in the workplace forum where the participants were encouraged to share what was important to them and contribute to the discussions. She adds that thanks to these forums, she had the chance to get to know her colleagues much better, especially regarding their social backgrounds, i.e. ethnic origin, the socio-economic status of their families, their religion, sexual orientation, or political affiliation. Therefore, the process that started with the protests caused her not only to develop a political consciousness, but also to transform her approach to social relations. She describes how her point of view widened dramatically:

> You were a person with blinders on, then suddenly your vision became wider. You started to understand other people and even this attempt to understand is a very good thing. You listened to everyone. (26, employed, no political affiliation and member of a local initiative)

Listening to and understanding each other is underlined as the basis of this new way of relating to others that was discovered thanks to the protests. The emphasis on the very simple actions of listening and understanding is interesting as it also reveals a critical view of the way people used to relate to others. To put it differently, attentive listening and empathetic understanding are seen as a novelty, or something that was rediscovered during the protests, because a lack of communication and contact between people, especially those from different backgrounds, had been the norm for a long time. In her account, the interviewee engages in self-criticism for having been someone with "blinders on"—in other words, she was somehow mind blind, and had low awareness of broader society. A participant who became an active member of a right-to-the-city group after the protests shares how she realized that her attitude toward many groups in society had previously been discriminatory:

> Coming together with those people, and trying to understand what they want and things like that... I certainly do not want to fetishize or romanticize the Gezi Protests but Gezi showed me lots of things. I had been homophobic, I had been a racist person, I had been discriminatory on many issues, I became aware of that... I realized that I had indeed a very discriminatory attitude toward Kurdish people. When I talked and listened to some of these people, I became aware that what I had been doing didn't make sense. (24, student, right-to-the-city activist)

Illustrating the interviewees' changing perception of Kurdish people, this statement clearly depicts how open and empathetic ways of relating to different groups in society—free from prejudice and exclusionary and discriminatory mainstream discourses—creates an opportunity to overcome or at least renegotiate the divisions and polarizations between people with different socio-political backgrounds. Moreover, changing their modes of relating to others and the newly developed awareness of the value of collective action eventually affected young people's relations with the broader public. In that sense, it seems that the protest experience has encouraged young people to move from an individualized way of life toward a stronger sense of community and the common good. An activist shares how she sees her protest experience in Gezi as being related to her motivation to take part in the solidarity activities organized following the Soma Mine Disaster in May 2014, a year after the Gezi Protests:

> After Gezi, I wanted to listen to people. Everyone has a different story. For example, I go to the protests and when I see a faction I have never seen before, I go and ask about their issues and goals. I have started to see everyone's validity. I started to care about other people's grievances, hopes, and struggles… For example, if Gezi had not

> happened, I would not have gone to Soma.[31] You can call it encouragement. I went there. I think that my motivation for listening, seeing, understanding people and doing something for them comes from Gezi. (22, student, Kurdish and feminist activist)

Therefore, it is plausible to argue that this transformation, which enables closer and more real contacts with those previously labeled the "others" in society, is to a great extent the result of the experience of collectivity young people gained in the Gezi Protests. Moreover, this experience of collectivity and its implications in terms of changing young people's ways of relating to each other and society in general have also been reflected in the new politics that emerged from the Gezi Protests. For example, young people frequently highlight relations between individuals, and their situational awareness in society as areas in which the political is built, shaped, and practiced.

### 4.4.2 Increasing Political Interest and Consciousness

A high level of awareness of one's social environment goes hand in hand with consciousness in the political sphere. An interviewee without political affiliation relates how she and her friends became more interested in politics in the aftermath of the protests:

> I do not remember that we ever talked about the government or a protest in high school or in the first years of university. But after the Gezi Protests, whenever we came together, we started to talk about these things, everyone's thoughts regarding these issues started to form. Many of my friends started to go to solidarity forums; however I am still a bit reluctant. But I do feel that something has changed. (24, student, no political affiliation)

In many of the interviews, young people emphasized how they lost trust in the mainstream media throughout the protests. The broadcasting of a documentary about penguins on CNN Turk while a harsh police intervention was taking place in the park marked a turning point at which the media were completely discredited in the eyes of the protesters. Young people emphasized how they became

[31] The interviewee mentions her participation in the solidarity events in Soma following the Soma Mine Disaster, which was one of the biggest industrial disasters in Turkey that left 301 miner dead. For a related news article, see: https://www.reuters.com/article/us-turkey-mine/turkish-mine-disaster-town-under-lockdown-as-death-toll-rises-to-301-idUSBREA4C0KO20140517 (accessed on April 10, 2018).

skeptical of the Turkish news media after the main news outlets had been ignoring the protests for a long time. This distrust was also coupled with their rising political consciousness interest and brought about a considerable change in young people's relationship with the media. The quotation below illustrates the extent of this loss of trust in the mainstream media:

> I had already known that the media was distorting some of the news or spotlighting some while ignoring others, but I had never witnessed such a thing... I could not expect that they would censor and turn a blind eye to such an event taking place in the city center. That is to say, I lost my trust, it became clear to me that I should have never trusted the media in the first place. (24, student, no political affiliation)

After a long period of protesting on the streets and camping out in the park, protesting was no longer an unfamiliar practice for young people. Considering that Gezi was the first protest experience for many, it is especially important that young people became more accustomed to protesting on the street and putting up resistance. An environmental activist describes this as "knowledge of protesting":

> For many people, it was the first time they were taking part in a protest... As I said, people developed a consciousness, a knowledge of protesting, and were not afraid of the protest. (23, student, environmental activist)

It is significant that the protest experience has changed, developed, and transformed young participants' relationship with the political sphere. The experience of collective living in the park in particular, and the Gezi spirit in general, seem to have set in motion a transformation of young people's views on the possibility of political change and their individual and collective political agency. A right-to-the-city activist underlines that what the Gezi Protests taught the participants was, in essence, awareness of the need to defend their rights. Going through a process in which they claimed that a public green area should stay as it is, fought for it, and eventually won, showed the protesters that they could build their future struggles on this success:

> For me, the Gezi spirit is about the freedom to exercise your rights, human rights. We got our rights in the end. It is also about the awareness of the importance of green spaces in the city but the most important thing is people's successful struggle for their rights. They said that this green space was theirs by right and that they would take it, and then they took it. The protests left behind lots of things, but the most important thing is that. (23, student and employed, right-to-the-city activist)

This kind of learning also deconstructs the pessimistic perspective according to which political change in Turkey is impossible, an outlook that this generation inherited from their parents who developed it in their adolescence as a result of their traumatic experiences under military rule.[32] The experience of success in struggle for political rights has shown them that doing something in the political sphere is not necessarily futile, and that political change is possible if it is desired and worked toward. In addition to a political consciousness acquired through the realization of one's rights as a citizen faced with the state and restrictive government policies, the protest experience also introduced young people to new spaces and forms of doing politics. As discussed in the previous section, workplace forums, neighborhood forums, and occupation houses are examples of new spaces that emerged as an extension of the protests, in which the participants could discuss social issues and elevate their political consciousness. An activist who is also a member of a neighborhood solidarity initiative accentuates the politicization process that came after the protests in these newly emerging spaces:

> What politicized people was not Gezi or the 15 days in the park, because that was like a dream. The protest changed and became a park forum, then neighborhood forums and then occupation houses. This happened gradually and the process itself was what really politicized people. (23, employed, right-to-the-city activist and member of neighborhood solidarity initiative)

This shift in relations with politics, along with the nascent awareness of their basic rights, prompted young people to act, guided by their increased interest in social issues and politics. An activist who has been a member of a leftist organization for a long time and a member of the neighborhood solidarity initiative since the Gezi Protests shares her observations:

> I think that Gezi is about organizing struggle in every area of life. After Gezi, there was a protest about safety in public transportation. A young boy was beaten at the metro station, and the following day many people were on the streets to protest. We had already been organizing protests about public transportation beforehand but they never attracted that much attention… That time, it turned into a people's protest. It was a different version of Gezi… That is to say, I think that [Gezi] is continuing in many new ways. (25, employed, member of a leftist organization and a neighborhood solidarity initiative)

[32] The effect of generational transfer of political experiences on young people's relationship with politics is examined in Sect. 8.1. Criticizing, Redefining and Contextualizing the Apolitical Youth Discourse.

As this statement shows, the level and efficiency of participation in protests centered around different issues increased following the Gezi Protests. However, what is important here is not only the increase in participation, but also the transformation of young people's perceptions of politics and political action. Their involvement in the protests seems to have impacted their commitment to other political activities ("different versions of the Gezi Protests"); in this way, Gezi has retained its relevance in the political sphere long after the actual protests have ended.

### 4.4.3 The Individual Level: Networking, Making Decisions, Taking Action

The impact of protest participation on the individual is an especially interesting question. Young people's accounts reveal that they relate crucial life decisions to their experience of the protests. Many emphasize that the protests were a life-changing event that changed their aspirations and affected their life decisions and future plans. An environmental activist relates how the Gezi Protests, and especially her ongoing participation in some other protests in the aftermath of Gezi, caused her to contact and befriend the kind of people whom she was looking for:

> I now go to more different protests than I used to. I have made friends in these protests. In the last two years, I have probably met a number of people that I could otherwise not meet in 10 years... I have made friends who do not question what I do, where I sleep, or what I do with whom. That is nice and brings me joy, and it is something that I had been looking for my whole life, and I have found it now. I can be exactly who I am when I am with my friends. They don't judge me. (30, unemployed, environmental activist)

Another interviewee similarly narrates how she was surrounded with new friends in the process of the protests, and then left her family's house to live with her friends:

> I moved out of my parents' house after the Gezi protests. Gezi had a big influence on this decision. I met many new friends there, and I am also still in touch with many people from the solidarity circles and forums formed after the protests… These people were mainly the ones who encouraged me to make this decision. I had severe conflicts with my parents at home. I met my new friends in 2013, and then in summer 2014, I left my family home and moved in with them. (23, employed, neighborhood solidarity)

On the one hand, these stories articulate the social networking effect of the protests—unsurprisingly, membership in these newly formed networks brought about changes in their everyday lives. On the other hand, these aftereffects of Gezi in their lives give clues about young people's perceptions of the relationship between politics and everyday life. Their reflections on how the protests have transformed aspects of their everyday lives, such as whom they live or spend time with, illustrate that it is hard to expect young people to have a notion of the political that is distant from the micro politics of everyday life.

Another point raised by some of the interviewees is that the protest experience has made them rethink and question their interest in and dedication to the academic disciplines they were studying at the time. It is truly striking to find instances in which the protests apparently encouraged young people to make serious decisions about their academic life, such as changing study courses. In the account below, a student narrates how she made up her mind to leave her current major in engineering and started studying sociology after the protests:

> Most of my classmates were not interested [in the protests]. While instructors were canceling the final exams in social sciences departments, our department was doing nothing… I made a request, but the professor answered that he could not do anything for me since no other students had made a similar claim... Then I said to myself that I would not be able to be myself in this field, there would always be some restrictions… I decided to switch departments. Then, with a bit of motivation, I took the entrance exam once more, and so I am here again, and I am happy about this. (21, student, youth-work volunteer)

Furthermore, it is noticeable that many young people gravitate toward social sciences. This growing interest is connected with their increased interest in social problems and political issues as a result of their protest experience. An environmental activist who pursued a double degree in economics and English before the protests decided to quit economics after Gezi and focus on her studies at the English Department. Her perception of both study programs also changed dramatically:

> With the Gezi Protests, I realized that I do not want to study economics. When I became aware that the economics we learned at college is capitalist economics, all about using resources, minimizing cost, and maximizing profit, I had a revelation... I asked myself, what is the point of studying this? Then I quit after the protests... During that time, I also realized that the courses at my own department (English Department) were actually the ones I want to take. For example, the critical theory course, it was really nice, we read Marx, etc. Then I realized that I do not have to meet the expectations of others, I do not want that, I do not live for others, I will make my own way.

> This way my life will not be about working a corporate job, having a meteoric career, high income and a rented car, and traveling abroad for work. This was my mothers' dream, I would have definitely been this kind of person if she had it her way. I can resist the role my mother has laid out for me, I need to play the role I have created for myself. (24, student, environmental activist)

Similar to this participant's critical stance toward capitalism and the sway it holds over economics as an academic field, an interviewee working at a corporation shares how she has started to question the capitalistic system and, by extension, her job. At the time of the interview she was on the verge of a radical change, having made up her mind to leave her job in the near future and get a graduate degree in environmental science.

> I started to become interested in [environmental] issues again right after the protests, I started to read much more; I started to improve myself more. Then I said to myself: I am changing, which means I have been ready to change for a while now, and now I will, permanently. I am now working toward this. I will leave capitalism, I will quit this job, and the first thing I will then do is get a master's degree in environmental science.... (26, employed, no political affiliation, attendant of a workplace forum initiative)

Similar to moving out of one's family home, changing one's academic field or leaving one's job and going back to university are important decisions that show young people are attempting to take control of their lives. In this regard, it is relevant to make a connection between young people's protest experiences, and their motivation to express themselves and make changes in their personal lives. That is to say, taking action on social issues seems to be strongly related to being active and having agency in one's own life.

### 4.4.4 Critical Views, Disappointment, and Traumatic Experience

Analyzing the transformative impact of the Gezi Protest on both the individual and the social level, it is important to be aware of the fact that not all young participants' accounts praised and celebrated the effects of the Gezi Protests. Rather, some interviewees were highly critical of perceived exaggeration of the impact of the Gezi Protests. They thought the transformation in young people's lives and especially the change in their level of political engagement to be overblown by many. Underlining the different ways people experienced the protests, an environmentalist states that the Gezi Protests were far from creating a long-lasting and effective change for the participants:

> People had different experiences there. That is to say, a white-collar worker did not come [to the protests] and changed into something else. There was a very important slogan in the Gezi Park, "Pluralist Democracy, Not Majoritarian." However, many people were not drawn by this slogan, they were not even aware of it… Gezi did not cause a big change for these people, the apolitical ones are still apolitical. I think that Gezi was an act of opposition, but it did not offer an alternative, it was just a reaction, it was not a place that taught politics. It started as a reaction and then fizzled out, it was a good awakening for people, it is true, we were all there and supported it. But for me, it was not something like a miracle; I did not see it that way. (24, student, environmental activist)

Another interviewee takes a self-critical stance while looking back on the protests. She agrees that the young people had had high expectations, as though the protests would never end or would inevitably result in some radical change. Moreover, she especially highlights how the forums created potential for political change, but were short-lived. She also points out that the forum experience helped her and other members of her feminist group identify potential allies and collaborators:

> When the forums started, we had one at the university, but they all ended soon. We are all responsible for this… If the forums had continued, we could have been in a much better position. But I am also aware that they enabled us to see who is and is not oppositional at the university, who is on our side, with whom we can get together now if something happens and what kind of things we could do... Forums can make things much easier, this is one of the things that the Gezi Protests have given us. (23, student, feminist activist)

In addition to critical opinions, some interviewees expressed disappointment with the Gezi Protests. These narratives of disappointment appear mostly in the interviews conducted months after the protests, but not in the ones made immediately after. For example, the interview with the feminist activist above was conducted in April 2015, nearly two years after Gezi. As time passed, and the protesters' expectations were not met, they began to feel disappointed. An environmental activist weighs in:

> Many people went out into the streets, we hoped that many things would change, and because this did not happen, we are disappointed. However, I still have some hope, and I will continue. (24, student, environmental activist)

Similarly, another activist shares how her feelings of happiness turned into disappointment over the two years period following the protests:

> It has been two years, and during that time, I felt happy in Turkey for the first time in my life. I felt that I could not be that happy in another country, and that I should give this country a chance. But then I understood again that it was wrong to give a chance to this country and this is what I am feeling at the moment. (30, unemployed, environmental activist)

These accounts of disappointment and negative evaluations of protest experiences not only prove the diversity of young people's perspectives but also give a sense of the various meanings read into the protests and different expectations people had of them.

Finally, it is important to keep in mind that traumatic events make up a considerable part of young people's protest experiences. As can be expected, exposure to police violence had staggering long-term effects on the participants. Many who witnessed or suffered police brutality report panic and anxiety issues at mass gatherings, in chaotic environments, and during police interventions. One of the participants recalls how the festive atmosphere in the park, in which small children with their families were playing and having fun, fell apart with the explosion of a sound bomb:

> I worked with children at the protests during the period when people were coming to the park with their children. Then suddenly a sound bomb went off. I heard the women and children shouting, people were turning red as a result of tear gas. Everyone was running, children were falling down, I still cannot forget this scene, and I am shaken even now. Ever since that event, my hands and knees start to shake when there is a police intervention at a crowded event and it becomes really hard for me to gather myself. (23, employed, volunteer in a civil society organization working for youth)

An LGBT activist insightfully shares that the lessons in solidarity and collective action learned at the protests were overshadowed by the memories of the violence participants had been exposed to, along with their enthusiasm for political struggle. The Gezi spirit, as much as it filled young people with hope during the protests, seemed to lose its power as the losses mounted.

> Yes, there was strong solidarity there but on the other hand there were losses, people were injured, people died and they will never come back. Now I am back to my hopelessness. Ok, we are shouting here, we are together, etc. but those people will never come back and with this in mind, I cannot figure out how a good and happy world can be built on the memory of those people... (28, student, LGBT activist)

Police interventions made it clear to the protesters that participation in a protest involved risking oneself both physically and in a legal sense. Awareness of this risk, along with witnessing the violence suffered by others, sometimes resulted in hesitation to keep protesting or participate in new protests. A right-to-the-city activist explains the gradual drop in participation in the protests that followed the events at Gezi:

> People are scared, I know. They are seriously afraid of getting involved in what is happening, being blacklisted, having something bad happen to them, for example if a gas bomb explodes during a demonstration... Many fears and anxieties… This is why many people do not take part. (31, student, right-to-city activist)

At the same time, these traumatic experiences caused young people to develop an awareness of the implications of state power. As many of them encountered the police and the state for the first time as "illegal" protesters on the street, they got a first-hand lesson in the workings of the state, police, and even the media. A protester relates:

> We often saw stories in the news such as there were protests at Gezi Park and the protesters dispersed peacefully... We saw that it was not like that. There is the story, but also a different reality behind the scenes. All of these things have been a real challenge for us. We understood what the state was and how important this thing called authority was. For me, it filled me with fear but at the same time I learned how to overcome it... (21, student, volunteer in a civil society organization working with youth)

## 4.5 Conclusion

This chapter has analyzed young people's experiences of the Gezi Protests. Informed by the interviewees' accounts of the pre-protest period, the chapter looks into the background of the protests. The participants report being overwhelmed by unhappiness and a sense of being oppressed by the neoliberal policies and authoritarian interventions. The analysis then continues with the protesters' reasons and modes of participation in the protests. Young people point out that it was the police brutality faced by the initial protesters in the park that moved them to join the protests in great numbers. Furthermore, the decision to take part was usually far from planned and coordinated, although many people decided to do so simultaneously. The chapter then focuses on the two-week occupation period, in which young people practiced communal living in Gezi Park.

Participants' accounts suggest that the park was a site where political ideals were put in practice and a collective identity based on solidarity built. The phenomenon of the Gezi spirit has also been scrutinized in this section by taking a detailed look at its characteristics, i.e. empathetic and respectful ways of relating to each other and feelings of happiness and trust emerging as a consequence of the collective that formed around this spirit. Finally, the chapter examines the effects of the protest experience on young people's lives by taking a look at the way in which the participants' ways of relating to each other and to society in general were transformed, and how their political consciousness developed. This is followed by a discussion of the impact the protests had on young people's individual lives, and the chapter ends by studying their critical views, disappointment narratives, and traumatic experiences. What this chapter has demonstrated is that emotions and moral values had a significant effect on young people's participation in the protests as well as on the formation of the collective identity during the protests. Practices such as solidarity, sharing, fostering respect and togetherness, and taking care of others were the main building blocks of the collective identity. These features of the Gezi spirit are reflected in the transformation of everyday relations in the protesters' lives, and in the sense of community that continued in the aftermath of the protests.

## References


Atay, T. (2013). The clash of 'Nations' in Turkey: Reflections on the Gezi park incident. *Insight Turkey, 15*(3), 39–44.

Bee, C., & Stavroula, C. (2017). Youth activists and Occupygezi: Patterns of social change in public policy and in civic and political activism in Turkey. *Turkish Studies, 18*(1), 157–181.

Coşar, S., & Yeğenoğlu, M. (2011). New grounds for patriarchy in Turkey: Gender policy in the age of AKP. *South European Society and Politics, 16*(4), 555–573.

D'orsi, L. (2015). Crossing boundaries and reinventing futures: An ethnography of practices of dissent in Gezi park. In G. Koç & H. Aksu (Eds.), *Another brick in the barricade: The Gezi resistance and its aftermath* (pp. 16–33). Bremen: Wiener Verlag.

Erdoğan, E. (2015). Siyasal Psikoloji Siyasal Katılım Hakkında Ne Öğretebilir? Gezi Protestoları'na Katılanlar Üzerinden Bir Değerlendirme. *Marmara Üniversitesi Siyasal Bilimler Dergisi, 3*(5), 31–58.

Gümüş, P., & Yılmaz, V. (2015). Where Did Gezi come from? Exploring the links between youth political activism before and during the Gezi protests. In I. David & K. Toktamış (Eds.), *Everywhere Taksim: Sowing the seeds for a new Turkey at Gezi* (pp. 185–200). Amsterdam University Press.

Jasper, J. M. (1997). *The art of moral protest: Culture, biography and creativity in social movements*. Chicago and London: The University of Chicago Press.

Kaya, A. (2015). Islamisation of Turkey under the AKP rule: Empowering family, faith and charity. *South European Society and Politics, 20*(1), 47–69.

Kaya, A. (2017). Right to public space: Social movements and active citizenship in Turkey. *Research and Policy in Turkey, 2*(1), 1–9.

KONDA. (5. June. 2014). *Gezi report: Public perception of the "Gezi Protests", Who were the people at Gezi Park?*, Retrieved January 5, 2022, from https://konda.com.tr/wp-content/uploads/2017/03/KONDA_Gezi_Report.pdf.

Kuymulu, M. B. (2013). Reclaiming the right to the city: Reflections on the urban uprisings in Turkey. *City, 17*(3), 274–278.

Lüküslü, D. (2016). Creating a pious generation: Youth and education policies of the AKP in Turkey. *Southeastern European and Black Sea Studies, 16*(4), 637–649.

Navaro-Yashin, Y. (2013). Editorial: Breaking memory, spoiling memorization: The taksim protests in İstanbul. *Society for Cultural Anthropology,* Retrieved January 23, 2018, from https://culanth.org/fieldsights/editorial-breaking-memory-spoiling-memorization-the-taksim-protests-in-istanbul.

Örs, İR., & Turan, Ö. (2015). The Manner of contention: Pluralism at Gezi. *Philosophy and Social Criticism, 41*(4–5), 453–463.

Özkırımlı, U. (2014). Introduction. In U. Özkırımlı (Ed.), *The making of a protest movement in Turkey #occupygezi* (pp. 1–6). London: Palgrave Pivot.

Özuğurlu, M. (2011). The TEKEL resistance movement: Reminiscences on class struggle. *Capital & Class, 35*(2), 179–187.

Tuğal, C. (2013). 'Resistance everywhere': The Gezi revolt in the global perspective. *New Perspectives on Turkey, 49*, 147–162.

Türkmen, B. (2014). Gezi Direnişi ve Kadın Özneler. *Kültür ve İletişim, 34*, 11–35.

Uzun, B. (2014). Devlet, Makbul Gençler ve Genç Kadınlar Bağlamında Kızlı-Erkekli Ev Meselesi, [Male-female Housing Problem in the Context of the State, Proper Youth and Young Women in Turkey]. *Kültür ve Siyasette Feminist Yaklaşımlar, [Feminist Approaches in Culture and Politics]*, 22.

Yeldan, E. (2010). Tekel Workers Resistance: A Re-Awakening of the Proletariat in Turkey, Retrieved January 23, 2018, from http://yeldane.bilkent.edu.tr/TEKELresistance_Yeldan2010.pdf.

Yıldırım, U. (2013). Editorial – An uprising on the verge of comparison. *Society for Cultural Anthropology,* Retrieved January 23, 2018, from https://culanth.org/fieldsights/editorial-an-uprising-on-the-verge-of-comparison.

Yörük, E. (2014). The long summer of Turkey: The Gezi uprising and its historical roots. *The South Atlantic Quarterly, 113*(2), 419–426.

# Part II
# Youth as Experience

# Studying Youth in Social Context

# 5

This chapter examines the concept of youth, reviews recent approaches to studying young people's lives, and takes a closer look at how being young in Turkey is studied, as well as how young people's participation in the Gezi protests has been framed with reference to the characteristics and experiences of the post-1980 generation in Turkey. The chapter aims to explain how this book approaches the notion of youth theoretically, and it examines the ways in which being young has been understood in the field of youth studies. This sets the background for the empirical discussions in the next chapter that explores the study participants' reflections on what it means to be young. These two chapters together intend to draw a detailed picture of the current dynamics of being young in Turkey, and the way they are reflected in young people's participation in the Gezi protests.

The chapter starts with conceptual discussion of the notion of youth. It relies on sociological and anthropological perspectives, which influenced the way this work approaches that concept, to describe age as a constituent of social relationships and, relatedly, youth as a socially constructed category. The chapter continues with a review of analyses of young people's lives in the field of youth studies that focus on youth transitions and youth cultures in accordance with the dominant trends, and a discussion of the social generations approach as a recent theoretical development attempting to understand youth in today's world. The second part of the chapter zooms in on the circumstances and implications of being young in Turkey. By way of focusing on the country's recent history, it discusses the social and political atmosphere in which young people of this generation have grown up. Finally, the chapter looks at how the generational aspects of the phenomenon of youth in the context of the Gezi protests have been discussed in the academia, including some studies that pointed out manifestations of the characteristics of the post-1980 generation youth in several aspects of the protests.



P. Gümüş Mantu, *Redefining the Political: Youth Experiences of Collective Action in Turkey*, Kulturelle Figurationen: Artefakte, Praktiken, Fiktionen,
https://doi.org/10.1007/978-3-658-40565-6_5

## 5.1 Conceptualizing Youth

It is beyond doubt that the way researchers in social sciences have defined the notion of youth has had an impact on how they have treated young people as their subject matter. Therefore this research, too, must concern itself with the concept of youth. In alignment with the current critical discussions of the ways youth has been conceptualized in social sciences that question the applicability of universal definitions of youth with reference to age, this work does not aim to establish a concrete definition of the term. Rather, it scrutinizes the implications of the conceptualization of youth as a social phenomenon and illuminates the conceptual framework through which being young is analytically studied in relation to broader social processes.

While discussing the definition of youth, researchers refer to the historical and institutional roots of the concept as we know it. Describing a social group with common characteristics and experiences as "young people" is a rather recent phenomenon dating back to nineteenth-century Western societies. Changes in the social and political sphere such as the introduction of new laws preventing child labor and the separation of employment from the domestic sphere with the beginning of the industrial revolution, coupled with the developments in adult citizenship rights in the fields of education, franchise, employment, and housing, resulted in the emergence of young people as a distinct group (Cieslik and Simpson 2013, p. 3). Therefore, youth as an intermediate stage between childhood and adulthood, during which young people are mostly in education, free from the responsibilities of adulthood and relatively independent from their families, is a product of modernity. Moreover, youth as a social category has been a politically, ideologically, and symbolically loaded concept and has stood for many things at the same time: "for the terrors of the present, the errors of the past, the prospect of the future" (Comaroff and Comaroff 2005, p. 20). Thus, deciphering the social significance of the phenomenon of youth promises not only a deeper understanding of its content in a specific context but also points to a revelatory potential in terms of wider societal dynamics.

A considerable share of the literature on youth is based on the assumptions of developmental psychology about universal stages of development, identity formation, normative behavior, and the relationship between social and physical maturation (Wyn and White 1997, p. 8). Thus, these earlier approaches to youth and adolescence mostly describe youth as a universal biological stage of the life cycle that people transit through in similar ways as they proceed to adulthood. As scholarly interest in youth (and childhood) spread in the second half of the twentieth century across the humanities and social sciences, sociology in particular, scholars started to analyze young people's activities and attitudes with reference to theories of socialization and subcultures in addition to frameworks of developmental stages and social transitions. Thus, this period, that started with youth studies in the 1970s and was stimulated by research on young people's leisure activities, and social inequality, points to an emancipation from approaches characterized by essentialism (Chisholm and Deliyianni-Kouimtzis 2014, pp. xiv–xv).[1] In this, sociological perspectives paved the way for scrutinizing the implications of being young within a given set of social circumstances, as well as for discovering young people's ways of responding and relating to their social milieu. As a result, prescriptive approaches that evaluate whether young people's behavior is normatively fitting for their age, and label deviants as deficient or abnormal, have become less prominent. By contrast, there has been an increase in descriptive studies that treat youth neutrally as a social category of research.

As a corollary, discussions aiming to clarify the theoretical basis of the category of youth sprang up in the atmosphere created by the new ideas in the field. Questioning the dominant perception of young people as a homogenous and static group in society, emerging perspectives have conceptualized youth as an "age-related process" putting the emphasis not on "the inherent characteristics of young people themselves, but on the construction of youth through social processes (such as schooling, families or the labour market)" (Wyn and White 1997, pp. 8–9), and criticizing the categorical approaches to youth, which assume age as the defining characteristic of youth. Overlooking difference, process, and change, understanding young people's lives as reduced to an age category presents an ahistorical and static picture of youth (Wyn and White 1997, p. 13). On the contrary, as a relational concept, youth refers "to the social processes whereby age is socially constructed, institutionalized and controlled in historically and culturally specific ways" (Wyn and White 1997, p. 11). Therefore, the relational perspective opens up the space for studying young people as a group whose members have diverse experiences, as well as youth as a concept with different meanings in different circumstances.

[1] For more classical examples of youth studies in this period, see (Chisholm and Deliyianni-Kouimtzis 2014, pp. xiv–xxix).

### 5.1.1 Youth as a Social Relationship

The sociological inquiry into the concept of youth underlines that youth cannot be merely defined with reference to age or biological development; rather, being young points to a social positioning and is realized within social relations. Pierre Bourdieu, while discussing how a sociologist can deal with the issue of youth, frames division between young and old as a power relationship:

> … the logical division between young and old is also a question of power, of the division (in the sense of sharing-out) of powers. Classification by age (but also by sex and, of course, class…) always means imposing limits and producing an *order* to which each person must keep, keeping himself in his place. (Bourdieu 1984 tra. 1993, p. 94, emphasis in original)

Defining age as a power relationship paves the way for the analysis of inequalities, hegemonic relations, and instances of oppression in relation to the ways age-based categories like childhood, youth, and adulthood function in the construction of the social order. Thus, similar to questioning the "normalcy" of power relationships based on sex, critically investigating how age operates in society in taken-for-granted ways, uncovers how the notion of youth is constructed, and how social norms are in turn constructed in relation to age, especially to youth. In addition, Bourdieu underlines the irrelevance of defining a universally applicable age-based category of youth by giving the example of how being a student and being a working-class youth results in enormously different experiences. He states that:

> … merely talking about 'the young' as a social unit, a constituted group, with common interests, relating these interests to a biologically defined age is in itself an obvious manipulation. At the very least one ought to analyze the differences between categories of 'youth', or, to be brief, at least *two* types of 'youth'. For example, one could systematically compare the conditions of existence, the labor market, the time management, etc., of 'young people' who are already in work, and of adolescents of the same (biological) age who are students. On the one side, there are the constraints of the real economic universe, barely mitigated by family solidarity; on the other, the artificial universe of dependency, based on subsidies, with low-cost meals and accommodation, reduced prices in theatres and cinemas, and so on…
>
> In other words, it is an enormous abuse of language to use the same concept to subsume under the same term social universes that have practically nothing in common. (Bourdieu 1984 tra. 1993, p. 95, emphasis in original)

Bourdieu suggests that "'youth' is just a word," (Ibid, p. 94) underlining the inapplicability of any universal definition of youth since there are various different experiences of being young in different socio-economic and cultural contexts. Thus, a sociological understanding of youth necessitates contextualizing young people's experiences by taking different social factors into consideration. Analyzing young people's transitions to adulthood is only meaningful in relation to certain circumstances of social, political, and economic conditions (Wyn and White 1997, p. 15). In other words, being of a certain age turns out to be a completely different life experience in dissimilar settings. Studies on individuals' varying ways of going through the youth phase have been conducted mostly with reference to structural factors such as class, gender, and race. In addition to these, disability, sexual orientation, and spatial divisions have also been discussed as the determinants of young people's experiences (Furlong 2013, p. 24). It is beyond doubt that these circumstances that create divisions do not solely and specifically affect youth lives, but rather shape the very basis of society in general. However, it seems crucial to examine how those circumstances specifically influence youth and result in a variety of ways of being young, not only because they affect the future, but also on account of the implications they have for the youth phase in the present moment:

> …the effect of these processes will be revealed in the 'society of the future' as social divisions are entrenched. But they also have an impact on young people's lives *in the present,* as young people negotiate and engage with class relations, sexism and racism. (Wyn and White 1997, p. 149, emphasis in original)

This perspective, that sees youth as an age-based power relationship and emphasizes the importance of the contextual dimensions and structural divisions in shaping young people's experiences, serves as the basis for critical engagement with the concept of youth by explicating its standardizing and reductionist aspects.

Framed as a social construct and approached relationally, youth as a concept points to a relative discursive space that calls for continuous rediscovery and reinterpretation of the term in different contexts. Thinking back to Bourdieu's notion that "'youth' is just a word," therefore prompts the question as to what studying youth is good for, or, as Deborah Durham puts it: "If youth is both relational and culturally constructed, even possibly a social effect of power, how can we propose it as a general topic for investigation?" (Durham 2000, p. 115). Adopting an anthropological perspective in understanding youth in Africa, Durham defines it "less as a specific age group, or cohort, but as a 'social shifter.'" Originally a linguistic term, "shifter" refers not to absolute referentiality to a fixed context but to a relational, or indexical, context, as in the case of terms such as "here" or "us" (Durham 2000, p. 116). Thus, conceptualizing youth as a shifter makes it possible to see its revelatory potential in terms of the structure within which it exists:

> Shifters work metalinguistically, drawing attention to specific relations within a structure of relations, to the structure itself. This seems to be particularly the case with the mobilization of the idea of youth in social life. In recognizing, experiencing or disputing "youth" in everyday life, people draw attention to the ways relations are situated in field of power, knowledge, rights, notions of agency and personhood.... To imagine youth, and to imagine the concept relationally, is to imagine the grounds and forces of sociality. (Durham 2000, pp. 116–117)

Inspired by Durham's approach, this work concentrates on the metalinguistic aspect of the concept of youth and treats youth as a shifter that reflects broader structural dynamics. Therefore, looking closer at how young protest participants perceive and experience being young offers insights into the context and structures they live in. Within this framework, the interviewees' accounts of hardship, oppression, and obstacles are interpreted as articulations of the ongoing socio-economic and political processes in Turkey. Moreover, the way young people reflect on their participation in the protests and their political potential in general is analyzed as a manifestation of the ways in which politics, power, and age are perceived in the society. To put it more clearly, this work studies how broader social processes are mediated through the notion of youth, and therefore how the experience of youth emerges within the dialogical and dynamic relationship between circumstances, young people's actions, and agency. Conceptualizing youth as a shifter makes this relational perspective possible and transforms youth into an analytical category reflecting complexities of society beyond its categorical boundaries.

### 5.1.2 Studying Young People—The Concept of Generation

The field of youth studies has long been divided into two different dominant approaches (the cultural and the transitional) to understanding young people's lives. While the cultural perspective concentrates on distinctions and complexities of cultural representations and responses by employing ethnographic and small-scale research focusing on local instances over a short period, the transitional perspective mostly focuses on the relationship between education and work and uses large-scale quantitative studies in different locations over a longer period of time (Furlong Woodman and Wyn 2011, p. 356). Historically, the cultural approach stems from the work on youth subcultures done by the Centre for Contemporary Cultural Studies (CCCS) at the University of Birmingham, UK, in the 1970s. The Centre's research mainly focused on subcultural styles as group-centered expressions of resistance to structural marginalization of the working class, and social change in post-war Britain. As a result of the discussions on postmodernity, risk and individualization coupled with internal critiques of the reductionist emphasis on subcultural styles, the cultural approach later broadened its focus to include private as well as public spaces and the cultural practices and identities of "ordinary" young people, not just those who were most visible (Geldens, Lincoln and Hodkinson 2011, p. 348). In contrast, the transitional perspective, which became a dominant trend in youth studies in the 1980s, is largely based on developmental approaches and sees a young person's life as a linear progression through institutional stages such as education, employment, marriage, and independent living (Wyn 2014, p. 7). Changing perspectives in youth transitions were reflected in changing metaphors used to characterize young people's passage through the institutional stages. In the 1980s, "trajectory" was the predominant metaphor used to describe youth transitions by emphasizing that transitional outcomes were conditioned by structural factors such as social class and cultural capital, and thus went beyond individual control. The popularization of postmodernist approaches and their emphasis on individual agency, caused "navigation" to emerge as a new metaphor underlining individual skills such as judgment, resilience, and life management skills (Furlong 2009, p. 2). Despite the development and transformation of both the cultural and the transitional perspective, the absence of productive scholarly dialogue between the two and its implications for the field of youth studies has recently been put on the agenda by many researchers (Furlong, Woodman, and Wyn 2011; Geldens, Lincoln and Hodkinson 2011). In this context, using the concept of "social generation" to understand young lives in contemporary world is seen as a theoretical stance acknowledging the historical-structural factors affecting young people's lives as well as their ways of negotiating with these factors, therefore their cultural practices and subjectivities are included in the analysis:

> Hence, a social generational approach requires the systematic analysis of the economic, political, social conditions impacting on young people. At the same time, because this concept rests on the notion that distinctive forms of consciousness are produced by changing social conditions, a generational approach takes account of the role of culture and subjectivities as forces for social change. (Furlong et al. 2011, p. 361)

Although generation is not a new concept in sociology of youth, it is possible to argue that its explanatory potential and analytical complexity have long been underestimated in youth studies.[2] The term gained popularity in mass media and popular literature as the use of labels such as Generation Y and Generation X, the Lost Generation, Millennials etc. became widespread. However, these designations have "served as a short-hand reference to describe the characteristics of successive cohorts of young people" (Wyn and Woodman 2006, p. 496), and have not lead to any further insights or analysis.

The sociological definition of the concept of generation has largely been built upon Karl Mannheim's seminal essay "The Sociological Problem of Generations" (Mannheim 1952). Mannheim defines generation as a "location," based on age groups and formed in the context of socio-historical processes. He argues that: "The unity of generations is constituted essentially by a similarity of location of a number of individuals within a social whole" (Ibid, p. 290). The concept of generation becomes an "actuality" for individuals born in a specific period, but collective exposure to socio-historical conditions of that period also plays an essential role. While the ways individuals experience these conditions or respond to the context they live in may be quite different, members of a generation still share the experience of living in the same historical and social reality. The differences in their relation to this broader context are the fundament on which generation units are formed:

> Youth experiencing the same concrete historical problems may be said to be part of the same actual generation; while those groups within the same actual generation which work up the material of their common experiences in different specific ways, constitute separate generation units. (Mannheim 1952, p. 304)

Mannheim goes on to discuss the potential for young people's transformation and the development of consciousness inherent in the social reproduction process, and he introduces the term "fresh contact," meaning "meeting something anew." Fresh contact implies a "changed relationship of distance from the object and a novel approach in assimilating, using, and developing proffered material" (Ibid, p. 293). Thus fresh contact of young people with the already existing cultural and social material in society opens up possibilities for new ways of relating and negotiating with it. In that sense, the social reproduction process "facilitates reevaluation of

[2] Jane Pilcher underlines several points sociology of generations concerns itself with that are fundamental to sociology as such, and argues that the social concept of generation did not attract the attention it deserves in British sociology. For more, see: (Pilcher 1994).

our inventory and teaches us both to forget that which is no longer useful and to covet that which has yet to be won" (Ibid, p. 294).

Thus, understanding youth through the lens of the concept of generation enables an analysis of how young people are located (in Mannheimian terms) within the larger historical socio-economic context, how they respond to the articulations of this location, as well as how their fresh contact with the dynamics of society opens up new possibilities of consciousness, reaction, negotiation, and action. Owing to this theoretical framework, how young people are exposed to common experiences in the current global socio-economic circumstances as well as how they react to these broader processes can be studied in a relational manner. Therefore, the social generations approach offers the means to trace structural and transitional processes with a strong emphasis on individual agency as a mediator between these processes in everyday life. That being the case, young people are taken as active subjects in shaping their generational characteristics, that is to say, in addition to social, economic, and political conditions, for understanding a generation "it is also important to understand the role that young people themselves play in constituting distinctive features of their generation" (Wyn and Woodman 2007, p. 500).

Following the social generations approach, this study will pay special attention to the local and global contextual factors that bear on young people collectively. Accordingly, the interviewees' reflections on being young are interpreted in relation to their common experience of growing up in Turkey in the post-1980 period, that is, the 1990s and 2000s. I look at how this generation's specific experiences in several areas, such as education, work, family, public space, politics or symbolic and cultural space, coconstruct "being young" for its members. It is within this framework that young protest participants' ways of relating to these generational features in the context of the Gezi protests are analyzed.

### 5.1.3 Being Young in Late Modernity: Some Reflections

Young people's lives have in recent years become increasingly characterized by inequality, insecurity, and uncertainty. Young people and children in different parts of the world experience different manifestations of similar hardships such as poor education, inability to remain in formal education, the widespread phenomenon of youth unemployment, as well as police harassment, corruption, poor housing options, and intergenerational tensions (Jeffrey 2011, p. 146). While states are getting progressively less supportive of young people via social and economic policies, the ways young people are oppressed institutionally are increasing in numbers and becoming more effective:

> In essence, monetarist policies have relied on the categorizing power of age as a primary mechanism for narrowing the fiscal responsibility of the state for young people's lives while at the same time hugely expanding the reach of monitoring, surveillance and control over both young people's lives and the institutions in which they spend their time… This process is far from coherent, but its overall effect is that being young became more costly (and the price of failure even more expensive). (Wyn and Woodman 2007, p. 499)

Ulrick Beck's concept of "risk society" (1992), with which he describes the historical transformation of industrial society into a new modernity characterized by the notion of risk, has been influential in the analysis of young people's lives in recent years. While modernity implies rationality, predictability, and the belief in the potential of scientific knowledge, in late modernity the world turned out to be a dangerous place in which its inhabitants are constantly confronted with risk (Furlong and Cartmel 2007, p. 3). Thus, uncertainty and continuous negotiation of several risks ranging from environmental threats to anxieties about unemployment become the defining features of life in the risk society. Even if "type, pattern, and media for the distribution of risks differ systematically from those of the distribution of wealth," risks are usually allocated in a "stratified or class-specific way" (Beck 1992, p. 35). Accordingly, social inequalities, in parallel with class structures and socio-economic inequalities, tend to persist in the risk society. What is distinct in terms of how people experience social inequalities is the process of individualization:

> The tendency is towards the emergence of individualized forms and conditions of existence, which compel people—for the sake of their own material survival—to make themselves their own planning and conduct of life. (Beck 1992, p. 88)

The process of individualization and subjective feelings of risk emerge as some of the defining features of young people's lives in late modernity. Throughout their transition from childhood to adulthood, while passing through institutional processes and aiming to secure their future lives, young people increasingly act with the assumption of "individual responsibility and accountability," coupled with increased levels of "vulnerability and lack of control" (Furlong and Cartmel 2007, p. 9). That is to say, although it is often ignored, individualization operates in conjunction with the structural processes. Youth researchers tend to argue that despite the manifestations of extensive social change, young people's experiences continue to be shaped by class and gender while growing up in the risk society (Furlong and Cartmel 2007, p. 8). Furlong underlines that the biographies of individual young people are formed within the boundaries of structural factors:

> Biographies may be negotiated, but such processes occur within the constraints of structures based on factors like class, gender and 'race'. Individualization is a structured process and exposure to risk is dependent on socio-economic location. (Furlong 2006, p. xvi)

However, this emphasis on the determining function of the structural processes does not exclude young people's agency in negotiating their way through these circumstances. Furthermore, spaces and experiences, which make young people feel stuck with the burden of individual responsibility and the structural limitations it imposes, emerge as the origins of youth agency. To put it differently, young people are not passive victims lost in these processes; rather, they are actively engaging with them and negotiating their position:

> Youth agency can only be apprehended by understanding how children and youth navigate plural, intersecting structures of power, including, for example, neoliberal economic change, governmental disciplinary regimes, and global hierarchies of educational capital. (Jeffrey 2012, p. 246)

These reflections on the current conditions of being young are certainly far from being able to explain the circumstances of young people's lives on a global scale.[3] Rather, this work takes the discussions of risk society, individualization, and young people's experiences of uncertainty and vulnerability within these processes as guiding conceptual points in the analysis of young people's experiences in Turkey. Informed by these theoretical positions, this research interprets young participants' reflections on being young by paying special attention to how they, as individuals, negotiate their way through broader institutional and contextual dynamics. Furthermore, it examines how these collective experiences of individualization and risk as well as feelings of insecurity and anxiety play a part in determining generational location and generational consciousness.

---

[3] Beyond offering all-encompassing and universally valid explanations, it should be noted that these theoretical discussions reflect the Western bias in the field of youth studies, where research is particularly focused on young people in Europe and the Global North. For critical perspectives on Eurocentric approaches in youth studies, see (Durham 2000) and (De Boeck and Honwana 2005).

## 5.2 Being Young in Turkey in the 2000s: Socio-Economic Circumstances and Political Atmosphere

This section looks closer at the conditions of being young in Turkey in the recent decades. This is done with a view to gaining insights into the social context in which the interviewees grew up. It also discusses the generational aspects of the Gezi protests, which helps contextualize the common interpretation of young people's participation in the Gezi protests as a generational experience.

Demet Lüküslü, in a study that analyzes the construction of youth as a social category in Turkey's modern history, argues that in spite of the changes in political tendencies and contextual dynamics, young people's existence in society before the 1980s had been shaped by "the myth of youth," which implied an idealized perception of young people having political agendas and representing the future of the country. Lüküslü's analysis shows that the post-1980 period points to a rupture in that myth, where young people are no longer perceived as having a political agenda and responsibility, but as showing apolitical tendencies, apathy, and individualistic attitude (Lüküslü 2009). Hakan Yücel and Demet Lüküslü argue that multiple myths of youth coexisted in Turkey in the 2000s. The desire and attempts to construct the category of youth and the practice of ascribing a political mission to young people definitely continue to exist, albeit in different ways. Along with their ideologies, practices, and priorities, different political projects produce different myths of youth (Yücel and Lüküslü 2013, p. 16). Therefore, the social and political construction of the category of youth in Turkey in the last decades can be understood by deciphering the processes of "creating the myth of a pious youth" (Lüküslü 2016). The emergence of pious youth has undoubtedly been strongly linked to the radical change of Turkey's political and social atmosphere that occurred in the early 2000s under the rule of the Justice and Development Party. Ayhan Kaya stresses the links between neoliberal social policies and the religious forms of governmentality in the context of Turkey, and, while emphasizing the changes in the discourses and policies on family, education, and charity, he asserts that Turkish society has undergone a radical transformation under "the joint influence of Islam and neoliberalism" which is reflected in many aspects of everyday life. He describes the process as follows:

> There has been a subtle Islamization of society and politics in everyday life through the debates on the headscarf issue, Imam Hatip schools,[4] faith communities and

[4] Schools for religious education included in the compulsory education system.

> Alevism, the rise of an Islamic bourgeoisie with its roots in Anatolian culture, the emergence of consumerist lifestyles, not only among the secular segments of Turkish society but also among Islamists, and finally, the weakening of legitimacy of the Turkish military as 'the guardian of national unity and laicist order'. These are all very important aspects of the ways in which Turkish society and politics have been radically transformed since 2002 under the joint influence of Islam and neoliberalism. (Kaya 2015, p. 63)

Lüküslü examines the manifestations of this momentous transformation with regard to political discourses on youth, the field of youth policies, as well as education policies during the third period of the rule of the Justice and Development Party between 2011 and 2014. The Prime Minister's speeches expressing his determination to "raise a pious youth,"[5] the introduction of optional religion classes at the secondary level of education and transformation of high schools into vocational and technical schools, which in reality turns many of them into religious schools, as well as the normalization of Islamic norms through youth policies and mass events for young people have been underlined as major manifestations of a process that has led to the emergence of a new myth of a pious youth in Turkey (Lüküslü 2016, pp. 641–644). Embedded in several spheres of everyday life and continuously reproduced in official discourses, the myth of pious youth has become highly effective in terms of shaping young people's lives. While some young people conform to the idea of pious youth, others, predominantly raised in secular families and/or have oppositional backgrounds, are marginalized and in some cases labeled as threats to society, which is reminiscent of how the previous myth of youth functioned (Yücel and Lüküslü 2013, p. 16). Thus this idealized image of the pious generation creates an exclusionary effect, especially for young people who do not agree with its political and ideological underpinnings but are still exposed to its all-encompassing effects in their social lives.

In summary it can be argued that the dominance of the myth of the pious generation and its deterministic and oppressive effects have impacted young people in Turkey. This is, however, not the only factor that has played a formative role in young people's lives. Socio-economic factors and inequalities, particularly prominent in work and education, have had an equally strong impact on the experience of being young in Turkey. Most importantly, the availability of family support seems to shape young people's lives to a great extent. Ayça Alemdaroğlu, on the basis of qualitative research with young people, asserts that even if young people in Turkey are provided more choices in the fields of social reproduction (such

[5] From the prime minister's speech, translated by Demet Lüküslü in her article (2016).

as education, work, and family life) than the previous generations, the economic and cultural limitations and opportunities of the environment they live in, and especially their families' economic and cultural capital, largely determine their identities and patterns of belonging (Alemdaroğlu 2005, p. 28). In a similar vein, employing a social policies perspective, Volkan Yılmaz defines the current structure of social and youth policies in Turkey using the "denied citizenship" model and underlines how the familialization trend embedded in this model is far from enabling young people to live their lives without relying on their family budgets and family connections (Yılmaz 2016, p. 11).

Given the central role of the family in helping young people navigate through education and work, structural inequalities tend to continue over generations. The way in which young people are included in the education system from elementary school to the end of their education is largely determined by their families' economic background. Recent research shows that one of the main obstacles to equal access to education in Turkey is low family income (Bakış et al. 2009, p. 22). The privatization of the education system and the state's gradual withdrawal from the responsibility of providing free and equal education for all have created a situation where families' economic status has become the main determinant for children and young people's educational opportunities (Ibid, p. 22). Moreover, young people's performance in the university entrance examinations, and therefore their options for university education, are also strongly influenced by the support they get from their families. Their experiences at the university level also prove that the university system is strongly characterized by inequalities in terms of educational and research capacity, prestige, quality, as well as social environment (Yücel and Lüküslü 2013, p. 19). As is the case with their peers around the world, for young people in Turkey, university education is far from guaranteeing a job and a reasonable standard of living in the future. White-collar unemployment has been increasing in the recent decades in Turkey (Bora et al. 2011). Yılmaz underlines that having a job for young people in Turkey may not automatically lead to economic independence and decent living conditions. For example, gainfully employed workers have longer working hours in Turkey compared to all other OECD countries (Yılmaz 2016, p. 9).

Young people's lives in Turkey in the 2000s were characterized by a political atmosphere formed at the confluence of Islamization and neoliberal policies that gave rise to the myth of pious youth. Political discourses, as well as youth and education policies were used effectively to create a pious generation, and consequently determined to a large extent the atmosphere in which young people lived. As expected, the political idealization of the religious youth created its Other and marginalized and excluded young people who did not fit that category. In addition to this, the conditions of being young in Turkey are largely embedded in the social inequalities reflected in education and job market insecurity. Because of the country's social policy regime, young people seem to be dependent on their families as their main source of support in finding their way through institutional processes which are remarkably exclusionary for young people coming from disadvantaged backgrounds.

### 5.2.1 Generational Characteristics of the Gezi Protests

The generational aspects of the Gezi protests have been acknowledged in the literature on the Gezi protests (Göle 2013, p. 8). The protests have been interpreted as the momentous event when a generation of previously discouraged and apathetic youth was radicalized and repoliticized (Kaya 2017, p. 3; David and Toktamış et al. 2015, p. 21). The visibility of the young generation, often referred to more specifically as the 90s generation, and particularly the use of humorous ways of protesting and references to popular cultural symbols have mostly been perceived as signs of the formation of a generational character of the protests. However, despite the widespread interest in this aspect of the protests, there is only a rather limited body of research on them that is informed by the sociological concept of generation.

Bahar Tanyaş focuses on the manifestations of the discourse of the "nonpolitical" generation, on the basis of qualitative research with young Gezi protestors, and argues that the protests have made it possible for young people to defend themselves and challenge the degrading effects of being perceived as apolitical. Thus, the Gezi protests point to a change in young people's ways of articulating their perception of others/authorities and have therefore had a notable effect on the young generation, making it different from previous generations (Tanyaş 2015, p. 43). Derya Fırat and Tuba Emiroğlu argue that young participants of the Gezi protests who were born in and after 1990 potentially constitute a new political generation, since for many of them Gezi was the first event in which they were organized as a political collective, which arguably makes it a milestone experience in their lives (Fırat and Emiroğlu 2014, pp. 398–399). Cihan Erdal draws attention to how the Gezi protests subverted the age-based hierarchical relationships, in leftist political organizations in particular and in the political sphere in general, and he looks at how they enabled the young generation to prove their political existence (Erdal 2017). On the basis of her exhaustive qualitative field research, Lüküslü discusses the characteristics of the post-1980 generation in Turkey through the prism of their practices and presence in the Gezi protests, and argues that the protests featured the characteristic language and humorous attitude of the post-1980 generation (Lüküslü 2017).

These studies, all of which rely on the sociological understanding of the generation to analyze young people's role in the protests make a valuable contribution to the analysis of the events at Gezi from a generational perspective and prove the momentousness of the protests for the generations that grew up in Turkey in the 2000s. Inspired by their effective use of the generational approach to study the Gezi protests, this work takes a closer look at how young people's protest experiences have influenced their perceptions of their generational identity, and especially the way they relate to the circumstances that make up the social background for their generational location. Within that framework, it studies the ways young people's individual experiences have translated into collective generational grievances and how the concept of generation mediated the process of the formation of a common.

## 5.3 Conclusion

In light of these theoretical and literature-based discussions, this research treats age as an analytical category and investigates how it shapes and organizes young people's social lives. The way young people experience being young in Turkey as well as the ways in which youth is constructed as a social entity are analyzed as reflections of broader social processes, with reference to the conceptualization of youth as a "social shifter" (Durham 2000). This study also specifically investigates the relationship between processes of individualization and collective experiences of social and structural change. The chapter discusses the notion of generational consciousness with a specific focus on the emergence of collective identity through practices of young people raised under the effects of individualization and scrutinizes the effect the Gezi protests have had on young people's ways of relating to the social processes around them, their individual agency, as well as the collective existence of their generation. This discussion is guided by the question of how young people perceive the collective phenomenon of generation, or, in other words, how they reflect on the contextual influences they are collectively exposed to when dealing with the hardships and limitations resulting from the overlap of socioeconomic and political circumstances as well as age-based power relations in society.

This work aims to illuminate the interplay between young people's participation in the protests on the one hand, and their experiences of and reflections on being young, as well as their generational location and consciousness on the other. In doing so, it seeks to explore in detail the unique importance of the Gezi protests with regard to the lives of the 2000s generation in Turkey, in contrast to short-cut understandings of the protests as an unexpected return of an apolitical generation on the political scene of the country.

## References


Alemdaroğlu, A. (2005). Bir İmkân Olarak Gençlik. *Birikim, 196,* 21–30.

Bakış, O., Levent, H., İnsel, A., & Polat, S. (2009, Febuary). *Türkiye'de Eğitime Erişimin Belirleyicileri.* ERG – Eğitim Reformu Girişimi. Retrieved January 5, 2022, from http://www.egitimreformugirisimi.org/wp-content/uploads/2017/03/TurkiyedeEgitimeErisiminBelirleyicileri-1.pdf.

Beck, U. (1992). *Risk society: Towards a new modernity* (Mark Ritter, Trans.). Los Angeles, New Delhi, Singapore, Washington DC: Sage Publications. (Original work published 1986)

Bora, T., Erdoğan, N., Üstün, İ., & Bora, A. (2011). *"Boşuna mı Okuduk" Türkiye'de Beyaz Yakalı İşsizliği.* Istanbul: İletişim Yayınları.

Bourdieu, P. (1993). Youth is Just a Word, In *Sociology in Question* (R. Nice, Trans.) (pp. 94–101). London, Thousand Oaks, New Delhi: Sage Publications. (Original work published 1984).

Chisholm, L., & Deliyianni-Kouimtzis, V. (2014). Introduction. In L. Chisholm (Ed.), *Changing landscapes for childhood and youth in Europe* (pp. xiv–xv). UK: Cambridge Scholars Publishing.

Cieslik, M., & Simpson, D. (2013). *Key Concepts in Youth Studies.* Los Angeles, London, New Delhi, Singapore, Washington DC: Sage Publications.

Comaroff, J., & Comaroff, J. (2005). Reflections on youth: From the past to the postcolony. In A. Honwana & F. de Boeck (Eds.), *Makers & breakers: Children & youth in postcolonial Africa* (pp. 19–30). Trenton N.J.: Africa World Press.

De Boeck, F., & Honwana, A. (2005). Children and youth in Africa: Agency, identity, place. In A. Honwana, & F. de Boeck (Eds.), *Makers and breakers. Children and youth in postcolonial Africa* (pp. 1–18). Trenton, N.J.: Africa World Press.

Durham, D. (2000). Youth and the social imagination in Africa: Introduction to parts 1 and 2. *Anthropological Quarterly, 73*(3), 113–120.

Erdal, C. (2017). Meşrunun Alacakaranlığı, Kuşağın Şafağı. In C. Erdal & D. Fırat (Eds.), *Devrimci Bir Pusula: Gezi* (pp. 132–145). Istanbul: Ayrıntı Yayınları.

Fırat, D., & Emiroğlu, T. (2014). "Gezi Kuşağı" mı Dediniz? Tarihsel Olay ve Toplumsal Kuşaklar. In D. Fırat (Ed.), *Sokağın Belleği* (pp. 371–400). Ankara: Dipnot Yayınları.

Furlong, A. (2006). Foreword. In C. Leccardi & E. Ruspini (Eds.), *A new youth? Young people, generations and family life* (pp. xv–xvii). London: Routledge.

Furlong, A. (2009). Revisiting transitional metaphors: Reproducing inequalities under the conditions of late modernity. *Journal of Education and Work, 22*(5), 343–353.

Furlong, A. (2013). *Youth studies: An introduction.* London and New York: Routledge.

Furlong, A., & Cartmel, F. (2007). *Young people and social change: New perspectives* (2nd edition). Berkshire: McGraw-Hill/Open University Press.
Furlong, A., Woodman, D., & Wyn, J. (2011). Changing times, changing perspectives: Reconciling 'transition' and 'cultural' perspectives on youth and young adulthood. *Journal of Sociology, 47,* 355–370.
Geldens, P., Lincoln, S., & Hodkinson, P. (2011). Youth identities, transitions, cultures. *Journal of Sociology, 47*(4), 347–353.
Göle, N. (2013). Gezi – Anatomy of a public square movement. *Insight Turkey, 15*(3), 7–14.
Jeffery, C. (2011). Geographies of children and youth III: Alchemists of the revolution? *Progress in Human Geography, 37*(1), 145–152.
Jeffrey, C. (2012). Geographies of children and youth II: Global youth agency. *Progress in Human Geography, 36*(2), 245–253.
Kaya, A. (2015). Islamisation of Turkey under the AKP rule: Empowering family, faith and charity. *South European Society and Politics, 20*(1), 47–69.
Kaya, A. (2017). Right to public space: Social movements and active citizenship in Turkey. *Research and Policy in Turkey, 2*(1), 1–9.
Lüküslü, D. (2009). *Türkiye'de "Gençlik Miti": 1980 Sonrası Türkiye Gençliği.* Istanbul: İletişim Yayınları.
Lüküslü, D. (2016). Creating a pious generation: Youth and education policies of the AKP in Turkey. *Southeastern European and Black Sea Studies, 16*(4), 637–649.
Lüküslü, D. (2017). Beklenmedik Buluşma: 1980 Sonrası Kuşağın Gündelik Hayat Dili ve Mizahı Direnişle Buluşuna. In C. Erdal & D. Fırat (Eds.), *Devrimci Bir Pusula: Gezi* (pp. 117–131). Istanbul: Ayrıntı Yayınları.
Mannheim, K. (1952). The problem of generations. In P. Kecskemeti (Ed.), *Essays on the sociology of knowledge* (pp. 276–320). London: Routledge & Kegan Paul.
Pilcher, J. (1994). Mannheim's sociology of generations: An undervalued legacy. *The British Journal of Sociology, 45*(3), 481–495.
Tanyaş, B. (2015). Gençler ve Politik Katılım: Gezi Parkı Eylemleri'nde 'Apolitik' Nesil. *Eleştirel Psikoloji Bülteni, 6,* 25–50.
David, I., & Toktamış, K. (2015). Introduction: Gezi in Retrospect. In I. David & K. Toktamış (Eds.), *Everywhere Taksim: Sowing the seeds for a new Turkey at Gezi* (pp. 15–25). Amsterdam: Amsterdam University Press.
Wyn, J. (2014). Conceptualizing transitions to adulthood. *New Directions for Adult and Continuing Education, 143,* Fall, 5–16.
Wyn, J., & White, R. (1997). *Rethinking youth.* London, Thousand Oaks, New Delhi: Sage Publications.
Wyn, J., & Woodman, D. (2006). Generation, youth, and social change in Australia. *Journal of Youth Studies, 9*(5), 495–514.
Yılmaz, V. (2016). Youth welfare policy in Turkey in comparative perspective: A case of 'Denied Youth Citizenship.' *Southeast European and Black Sea Studies, 17*(1), 41–55.
Yücel, H., & Lüküslü, D. (2013). 2000'li Yılları Gençlik Üzerinden Okumak. In H. Yücel & D. Lüküslü (Eds.), *Gençlik Halleri: 2000'li Yıllar Türkiye'sinde Genç Olmak* (pp. 9–24). Ankara: Efil Yayınevi.

# 6 The Meanings of Being Young: Potential, Limitations, Mission, and Action

This chapter investigates how young participants in the Gezi protests experience and perceive being young. Beyond the characteristics and traits commonly attributed to youth, the chapter aims to discover how being young is described in different contexts, and examines the way the meanings of being young relate to the experience of protest and the ways of relating to the political. The chapter first studies the interviewees' descriptions of youth as a period of potential and possibilities, before moving on to exploring young people's experiences of being young as a limitation and obstacle in various domains. This section covers a large cache of data concerning subthemes such as generational conflicts, societal pressure, gender, problems of the educational system, poor working conditions in the workplace, and pervasive feelings of anxiety about the future. The section includes young people's views on being young abroad, especially in Europe, in contrast to being young in Turkey in the oppressive political atmosphere in the country. This is followed by a closer look at how age relations function as a division through which young interviewees' experiences in social movements and political organizations are shaped and mediated. The final section examines how intergenerational relations were negotiated and the concept of youth revisited in the context of the Gezi protests.

## 6.1 Being Young: Positive Potential and Capabilities

The interviewees' accounts of being young mostly concentrated on the limitations and structural inequalities that turned this period of life into a difficult experience for them. Concerning this point, they mostly described their inability to experience being young "as it should be." The discussion below looks at how



P. Gümüş Mantu, *Redefining the Political: Youth Experiences of Collective Action in Turkey*, Kulturelle Figurationen: Artefakte, Praktiken, Fiktionen,
https://doi.org/10.1007/978-3-658-40565-6_6

they discussed these hardships and how they lived through them, but also examines the cases in which the interviewees framed being young as potential, in the sense that youth brings many possibilities and opportunities, and analyzes how the interviewees adopted and internalized the universally assumed characteristics of youth such as being energetic, dynamic, powerful, and care-free, as well as how these attributions acquired a meaning and function in their lives.

One study participant depicts youth as a good stage of life in which one has not yet started to be concerned about one's physical health and well-being but is at the same time psychologically capable of dealing with life's problems.

> It is a period when you do not worry about your body... You take everything seriously but at the same time you are psychologically strong enough to overcome everything easily... When you fall down, you can just laugh it off and share [the anecdote] with fifteen people... Everything is more joyous, funny, and seems worthy of sharing with others, I think... (25, no political affiliation, unemployed)

This account relies on the universally recognized image of youth. However, it is crucially important to contextualize this conception of youth by asking how experiencing youth in this way becomes possible. The quotation above is by a young person with undergraduate and graduate degrees from prestigious Turkish universities who had been looking for a job for a while at the time of the interview but did not complain about being unemployed thanks to ongoing family support. Moreover, she was able to afford to travel domestically and abroad often and had quite a lively social life in Istanbul, which is to say that she was able to live her youth according to the universal blueprint. It is plausible to argue that the propensity for defining youth by referring mainly to the positive potential inherent in it and the opportunities it provides is contingent on one's family situation, socio-economic status, cultural and political circumstances, and possibly many other factors. Ayça Alemdaroğlu, in her piece on being young in Turkey, argues that young people from different socio-economic backgrounds have a consensus on youth being defined as strolling around, doing as you please, going where you please. However, many of them argue that although they are young in terms of age, they do not feel young due to the economic, social, and cultural constraints they live under. Discussing young people's struggles for economic independence contextualized within the specific socio-economic, cultural, and political conditions, Alemdaroğlu insightfully defines "youth as an opportunity" (Alemdaroğlu 2005). Being young in its universal meaning is interwoven with the structural conditions in which one is going through the phase of youth.

Another interviewee relates being young with potential for and openness to new experiences and learning through trial and error. Thanks to the physical endurance of youth and the flexibility to adapt to changing conditions, this phase of life is thought to allow young people to do things they cannot do in the earlier or later stages of their lives.

> In simple terms, I see youth as a period in which one is free to try things out. It is such a nice energy, I think. I can go without sleep to achieve my goals, I can tolerate being short of money or being far from someone. (23, youth work volunteer, employed)

This interviewee was at the time of the interview very active as a volunteer in youth work, had a degree from a good university, and was working various part-time jobs to be able to live on her own. With this in mind, her description of youth with a special emphasis on new experiences and the willingness to give up even basic needs such as sleep for the sake of achieving what one wants to achieve is telling. As she is able to participate actively in many areas of life, this young participant perceives being young as having the power, potential, and ability to discover the world and realize ones' plans and ambitions.

A feminist activist who was about to graduate and had recently started to look for jobs sees being young as being free from adult responsibilities, especially with respect to financial matters. In line with the universal conception of youth, she considers being a student and having no other responsibilities important features of this stage of life.

> Being young for me is not having any financial concerns, your parents are still taking care of you, you are just studying and do not have any other responsibilities and you can do what you want. I myself still feel young, but as my graduation approaches, these thoughts are changing. (23, student, feminist activist)

This participant describes youth as partial independence from the realities of the adult world, in other words, a comfort zone free from financial and related concerns, and this perception certainly arises from her own experience of being young as a student, coupled with the universal image of young people as students.

These three examples, in which young people emphasize both the physical and mental potential, power, flexibility, sociality, openness to new experiences, and freedom from financial and other responsibilities, do not bring anything new to the table—all of these attributes have long been universally related to youth. In order to explain this it is useful to rely once again on Pierre Bourdieu's discussion of the different life experiences of young people from different socio-economic backgrounds. Since these universally accepted characteristics of being young mostly describe experiences of students supported by their relatively well-off families, then "merely talking about 'the young' as a social unit, a constituted group, with common interests, relating these interests to a biologically defined age" turns out to be a "manipulation" (Bourdieu 1993, p. 95). A critical understanding of these universal and stereotypical conceptions does not mean ignoring the existence of these features in young people's everyday lives. Rather, it is valuable to understand the context in which these features acquire meaning and purpose in young people's lives. The above discussion shows that the potentials and capabilities of young people that this stereotypical idea of youth stresses are not innate and do not come naturally with a specific age; rather, scrutinizing these assumed features of youth in context reveals their contingency on the circumstances in which young people experience being young.

## 6.2 Being Young: Hierarchies, Limitations, and Obstacles

This part of the chapter focuses on the interviewees' descriptions of being young in which they refer to difficulties related to their age and the obstacles and oppression they face in society. Conflicts with older generations, social pressures felt in multiple areas of everyday life, and gender-based power relations are examined in the first sub-section, which is followed by discussion of interviewee accounts focusing on their concerns about their future lives and careers, and an examination of how young people perceive their path through education towards work life. The third sub-section looks at what the interviewees' comparisons of being young in Turkey and in Europe reveal in terms of the conditions of how youth is experienced in the country.

### 6.2.1 Generational Conflicts, Social Pressure, and Gender

***Generational Conflicts***

As discussed in the previous chapter, the concept of generation is key to understanding what young people have in common with their peers in terms of the social and historical context they live in. In a similar vein, generational conflicts between successive generations can be understood as the discrepancy in the way members of different generations perceive their environment and society in general. In that sense, scrutinizing generational conflicts is vitally important as a means of making sense of the social construction of youth, by revealing how being young is currently conceived of and shaped in relation to other generations in a specific context. Moreover, with the definition of youth as a set of age-based power relations in mind, looking closer at generational conflicts provides insights into the intricacies of the power relations between generations in society. More specifically, how young people and the older generations see each other given these contradictions that exist between generations is worth analyzing in order to understand how contemporary definitions of youth and adulthood are created in the present social context.

The accounts examined reveal that young people mostly think that the older generations are highly critical of the younger ones and see them as inexperienced and inadequate. Young people report that, they are not treated as complete individuals in interactions with older generations, their parents relatives, and others, and that their opinions, experiences, and agency are dismissed or denied. One of the interviewees recalls how she felt dismissed, especially in her relationship with her family during her high school and university years:

> Of course, high school and university years are the ones in which the familial pressure is at its peak. Your words are not taken seriously. You are not accepted as an individual in society. In high school, you had already started to see yourself as an individual long ago, but it was like a classical adolescence period for me full of conflicts. (23, graduate student and working, oppositional Muslim)

This way of seeing young people as immature and incomplete is obviously related to the conceptualization of youth as a transition to adulthood. Accordingly, since being young is just the path towards being an adult, it is by definition impossible for a young person to be accepted as a complete, full-fledged individual in society. This perspective that sees youth as a deficiency is also evident in older generations' attitudes to undervaluing young people's life experiences and their potentials for change. A feminist activist describes how older people underestimate young people's opinions on account of their perceived inexperience:

> Being young in Turkey is difficult in a sense, because your opinions are not valued. It is like, "what did you live, what did you see?" If you have an idea about something, you can face this kind of attitude continuously. You can come across this point of view in people who are more experienced than you. (23, student, feminist activist)

Another activist tells how older generations underrated young people's potential for action and transformation on the basis of a biased comparison of their youth and young people's current situation with respect to political activities:

> People from the generation that came of age around the period from 1968 to 1970 like to make a distinction between them and us, and according to this, I am in the youth group as determined by them... It is always like... "We did not succeed, therefore neither will you, nothing will happen, this country will not change..." This opinion makes me really angry because our youth and their youth are not the same. (23, student, LGBT activist)

This approach to young people's political agency not only represents a pessimistic viewpoint but is also far from sensing the zeitgeist, in other words, the changing conditions and forms of political action between these generations. However, this view is also derived from the way the youth of the 1960s and 70s experienced political action and politics. As Lüküslü suggests, still dedicated to their mission as "the vanguard of the state," the youth of the 1960s and 70s sees it as its duty to fight the "illegitimate" regime, in other words the formal government. Distinct from the preceeding generations, "youth now became a "rebel" fighting against the system, later to be coined "anarchists" or "terrorists" by Turkish politicians or the press. This is a generation that opposes the values both of their families and of society" (Lüküslü 2005, p. 33). Undoubtedly, the way that the 70s youth generation perceived political action and change was shaped by their "rebellious" and "revolutionary" experiences that were characterized mostly by activists' commitment to traditional political organizations and taking part in violent street clashes. Thus their notions of the political and their ideas about the spaces of politics differ from contemporary youth's understandings of politics and political action that are mostly realized in everyday life, on the individual level, and in micro spaces. Framing older generations' pessimist and undervaluing attitude towards younger generations' potential for political action and change against this background is required in order to develop a well-rounded understanding of intergenerational relations.

***Social Pressure***

Participants of this study report that they are subjected to certain social pressures in their everyday lives as young people. They feel that older people mostly relate to them through expectations and responsibilities, instead of giving them a voice or advice on how to deal with their problems. Therefore, they believe that the way young people are seen and treated in society points to an insincere attitude that is extremely demanding but not supportive in return. An interviewee, who has very reflexive views on being young and the image of youth in society thanks to her voluntary engagement in youth work, criticizes the way young people are viewed in society:

> There is a double standard to society's perception of youth. They do not see young people as an entity that has a voice or is capable of doing and knowing something, but are always placing a burden of responsibilities on them. And it's not responsibility in its positive sense, like "you can do this well," but like "you have to offer your seat at the bus to the elders." While they do not value (young people) much, they still constantly expect something. (21, student, youth work volunteer)

Furthermore, some of the interviewees point out that they are exposed to stereotyping and judgmental treatment in the society. These complaints reveal that young people do not feel respected in society the way they are. On that point, it can be argued that older generations need a more empathetic and tolerant understanding of young people's behavior and lifestyle. An interviewee relates how a young person listening to music on their earphones catches critique and disappointment:

> For example, I have heard this many times: "What can you expect from a person who is listening to music on their earphones all day?" Is a person listening to music on their earphones all day really useless? They are making symbolizations for everything, they are segregating, and stereotyping. (23, working, youth work volunteer)

This anecdote points to a more general discussion on the way young people spend their leisure time and their resources. Whereas parents and social commentators usually have clear ideas of the content of healthy and constructive use of leisure time, young people very often spend their free time hanging around with friends, watching TV, or playing online games, surfing websites, or communicating via social media (Furlong 2013, p. 147). These various ways of "doing nothing"[1] are usually perceived by the general public as "killing time" and therefore useless, sometimes even harmful, yet studies on youth cultural practices reveal that they are crucially important for young people and provide them the time and space to reflect on their experiences (Ibid, p. 149). These cultural practices may also create a generation-specific vocabulary of communication, for example, based on the widely understood lingo of online games or popular drama series. Consequently, as was the case in the Gezi protests, this sort of shared vocabulary reflected in the chants, street writings, and social media can function as an alternative way through which young people express themselves and relate to politics, thereby contributing to the formation of the "youthful" character of the protests.[2]

---

[1] For a discussion of young people and "doing nothing," see Abbott-Chapman and Robinson (2009).

[2] The notion of "youthfulness" will be discussed in the fourth section to this chapter.

***Gender***

A considerable amount of social pressure young people are facing is related to their gender. Not surprisingly, young women's accounts reveal that they are increasingly subjected to criticism, as well as controlling and oppressive practices in society, and that this takes many different forms. A female participant narrates how she is feeling under pressure as a young woman and interprets this as her being unable to live her youth as it should be:

> For example, when I am hanging out with my friends, I do not want somebody to gossip about me for entering a house with a man. I want to drink alcohol in peace; I do not want people to judge me. This is the kind of youth I want. But social pressure is massive in Turkey, and I feel that I cannot live out my youth to the fullest. (24, student, no political affiliation)

On the basis of their qualitative fieldwork with house girls (*ev kızı*—young women who are neither in education nor in employment, who live with their families, and are usually waiting to get married), Kezban Çelik and Demet Lüküslü instigate a thought-provoking discussion on youth and gender in Turkey. One of the most important points of their research is that even if house girls are young according to their biological age, due to the conservative values of their social environment they think that they do not live like other young people, experiencing a "limited and restricted youth" instead. In contrast to "typical" young people, house girls are supposed to stay at home and do housework (Çelik and Lüküslü 2010, p. 38). What is striking is that the interviewee quoted above also reports that she does not feel that she is experiencing her youth fully. Being a student and coming from a relatively well-off family, her social background and everyday life practices have little in common with the lifeworlds of the house girls—for instance, from her account we understand that she is able to go out with her friends, even drink outside etc. However, she also feels under pressure, especially in the public sphere, as a result of the gendered and discriminatory views and treatment of young women in society. It bears stressing that young women commonly face conservative social pressure, albeit to different degrees and in different ways, which severely affects the way they experience being young and being women at the same time. The same interviewee goes on to relate how women in higher education, usually seen as more independent from their families compared to house girls, similarly have to contend with their families' restrictive attitudes and prejudices. Even if they are not expected to stay at home, they are continuously reminded by others around them of the need to make professional and career choices that leave room for domestic work and family duties. That is to say they are urged not to neglect their traditional position and function as women in society. The participant highlights how her family affects her important life decisions:

> For example, in my family, if I say that I want to be a surgeon, everybody overreacts. "How will you look after your child? How will you take care of your house? A woman cannot do this, it is so stressful..." There are so many generalizations. I am really sad about this. (24, student, no political affiliation)

Familial pressure, in other words, the family's decisive influence on one's life seems to be one of the most important characteristics of youth, especially for young women. Becoming independent of one's family's influence either by moving out, finding a job, or getting married, is mostly seen as one of the basic stages of transition from youth to adulthood, and these transition stages therefore mark important milestones in young people's lives. For young women, "standing on their own feet" and therefore being more independent is neither an easy nor a short process as the interviewee tells in these words:

> It is too difficult to become detached from your family in Turkey. That is to say, putting a distance between yourself and your family and starting to stand on your own feet. If you are a woman and if you are lucky, this may be possible if you graduate from university and find a job. The rents are too high to rent a house on a single salary. (24, student, environmental activist)

This opinion also illustrates how difficult it is for young people to gain access to basic rights, such as housing, even if they have a job. As discussed in the previous chapter, Volkan Yılmaz argues that the social citizenship dimension of Turkey's youth welfare citizenship model is weak and characterized by familialization. Young people are mostly dependent on their families' support in basic areas such as income, health, and accommodation. Therefore, the welfare regime in Turkey does not treat young people as adults but as children in need of family support regardless of their age (Yılmaz 2016, p. 4). In a similar vein, Laden Yurttagüler examines young people's situation in terms of civic rights in Turkey and how free they are to make their own choices, that is to say, their autonomy. As a consequence of the lack and the low quality of social services offered to young people, Yurttagüler underlines the family's role as a "safety net" in the fields of education, health, and accommodation. What is striking here is that "the attitude the families exhibit in meeting those needs determines the extent of young people's autonomy" (Yurttagüler 2014, pp. 33–34). With this in mind, young people's, and especially young women's, accounts of feeling constrained by the family are hardly unexpected, and they seem to be logical consequences of structural problems. In a similar sense, their desires of having "a life of their own," in the words of Ulrich Beck and Elisabeth Beck-Gernsheim (Beck and Beck-Gernsheim 2002), have to do with their liberation from being financially dependent on their families and constantly subjected to control and pressures.

Young women report uneasiness and insecurity in the public space as well, stemming from the social pressures they face, coupled with the threats of sexual harassment and violence against women. The quotation below is revealing in terms of how young women feel in public spaces and how they make adjustments to cope. Even though the interviewee is from a well-off family and spending time in the so-called safe parts of the city, she emphasizes that she is not free from this necessity to "protect herself" as a woman by being careful about her body, clothes, and gestures, in fact the totality of her presence in the public space. She explains that:

> Being a young woman in Istanbul means having too many bad experiences. Even if you come from a very sheltered background, I think that makes no difference, it's not like it happens to some people and not to others. I have had such bad experiences, and I hope I won't have any worse ones. It means constant anxiety, it means that you are constantly covering your body, walking fast, sulking. You have to know the limits and strategies. It means knowing how short your shorts can be. You cannot ever relax that you see European young people do. You have to be awake. You need to protect yourself. (25, no political affiliation, unemployed)

As the above discussion illustrates, gender shapes the experience of being young to a great extent in different ways. The overlap of gender and youth as power relations influencing the social reality results in young people facing multifarious forms of social pressure, restrictions, and obstacles and struggling for autonomy, independence, and individuality.

### 6.2.2 From Education to (Un)employment: Anxiety about the Future

Young people's accounts reveal that they generally perceive youth as a period in which they have to work hard to be able to make a future for themselves. Most of the interviewees report career choice concerns as early as in high school. Once they start college, they concentrate even more on choosing a career path and working towards building the foundations for it. As graduation approaches, the workload and stress of clearing a path for oneself seem to increase. Despite all these efforts, they eventually enter the job market and realize that it is not a level playing field and many have to face unemployment or accept jobs with poor working conditions. This trajectory is by no means the same for all young people. Throughout this transition from education to the job market, structural inequalities, especially socio-economic conditions that young people inherit, play a significant role. The quality of the education one receives as well as access to related networks that become essential during job search are exceptionally important factors that shape youth transitions, given the unfair nature of the market and the scarcity of jobs available.

This kind of excessive focus on the future seems to change and shape young people's relationship with the present. The interviewees reflect on being young in relation to constant anxiety about the future, as they expect it to be marked by oppression, overwork, and in some cases exploitation. The perception of youth as a transitory stage, as discussed in the previous chapter, crystallizes in this future-oriented way of experiencing youth. Framed as such, youth is only meaningful as the period in which one acquires skills and qualifications for a future career. In other words, adult life with a good job is the destination and youth is only the path leading to it.

A number of interviewees recall the preparations for university entrance exams as the point where they started to feel real pressure and anxiety, because how they scored in the exam would determine which, if any, of a large number of universities that vary in prestige and quality they would be able to attend. University entrance exams, a centralized aptitude test, were first administered in Turkey in 1974.[3] They have undergone several changes over the last decades in terms of their format, title, and content but success in the exams is still an entry requirement (although the minimal passing score differs across departments and universities).

Young people's experiences of entrance exams in Turkey have been the subject of a documentary film entitled *3 Hours* (Candan 2008), whose title refers to the duration of the exam. Documenting six young people's lives during the preparation stage, on the day of the exam, and afterwards, this film provides a detailed narrative of how young people from different socio-economic backgrounds and cultural contexts go through these processes. What the film effectively reveals is that structural divisions based on class and gender, or place of residence, e.g. living in a shanty town or near to the center of the city, affect the way prospective students experience the exams, their performances and therefore their future options, conveying at the same time the stress, depression, and isolation of the preparations.

[3] For a history of university entrance exams in Turkey (in Turkish), see the official webpage of the Student Selection and Placement Center (*Öğrenci Seçme ve Yerleştirme Merkezi*): http://www.osym.gov.tr/TR,8530/tarihsel-gelisme.html.

The interviewees' accounts of the university entrance exams were very similar to those in the documentary. A participant recalls his experience:

> Young people living in a way they do not want to live... They are becoming isolated and distant from social life because of the university exam. They might get fed up with this and rise up. (23, student, right to the city movement)

It seems that young people experience the preparations for the university entrance exams as a period of intense studying to score well enough to secure a place at a desired university/department or indeed any department or university. Such extreme workload and dedication result in a dramatic drop in free time, leisure, and family activities. But among adults, this suspension of normal life is considered as something integral to this unique period in a young person's life.[4] With this in mind, it is hardly surprising that young people associate this period with severe pressure and stress, and disruptions of everyday life routine. One interviewee highlights how young people's lives consist of a continuous series of tasks aimed at guaranteeing a future that, in essence, consists of a moderately prestigious job with an average income. But the entrance exam is just the beginning—what awaits in the years to come, both during their studies and after graduation, is the never-ending task of preparing themselves for the job market:

> For example, in Turkey, a young person is expected to be looking for jobs, should have already invested in his or her future, should have made her or his decisions about university education, should have spent a summer on an internships at a bank, should know from experience which sectors are the most lucrative, etc., it is so tiring... We get really exhausted in our youth, and on top of that comes the entrance exam, which is devastatingly stressful. (24, student, no political affiliation)

This account also lays bare how young people experience the transition from education to work as an extremely individualized process. The idea that making the right choices and putting maximum effort into one's future will eventually pay off and bring a good job and a decent life seems to obscure the impact of structural factors that shape young people's chances and quality of employment, such as the current dynamics of the labor market and the inequalities stemming from class or gender differences etc. Thus these structural experiences are "collectively individualized," framed as personal responsibility issues, that is to say, "your own life—your own failure" (Beck and Beck-Gernsheim 2002, p. 24), as people

[4] The documentary *3 Hours*, mentioned above, includes many examples of these sorts of attitudes to young people held by adults.

increasingly face exclusion, inequality and isolation. Unsurprisingly, young people feel exhausted and stressed out after high school, and then are expected to expand great effort to build a future for themselves. Families, relatives, and other adults around them usually keep a close eye on them during this transition, which only adds to the pressure. The statement below reveals that young people end up in a hamster wheel of trying to complete all the tasks their future is assumed to depend on, and inevitably give up on pursuing their own ideas and aspirations. Moreover, being under constant surveillance seems to intensify young people's anxiety about the future.

> They always ask: "Have you graduated already? What will you do for work? Have you started to look for jobs already? Will you go for a graduate degree? You had better go abroad..." I am still young, but I have started to feel severe pressure and I am really worried about what I will do. I have already found a job, but I still feel under pressure. Because in reality I do not want to do anything right now, since I have not decided what to do yet. I do not want to do something nonsensical just to kill time. What I understand about being young is you constantly think about your future, you plan for the future, you go to university for the future, you apply for an internship because that will pay off in the future, you attend a course because that will give you an edge in the future when you start your professional career. That is to say, you do not live your youth, you spend all your time investing in the future. (24, student, environmental activist)

Perceived as preparation for adult life, youth is thought to exist solely for the sake of working for the future as if the experience of being young now, in the present moment, were devoid of any value other than the time and energy required to improve oneself for the coming years. Ignoring young people's lives in the now as well as their feelings and ideas goes hand in hand with overlooking their rights and demands. While official political discourses often praise young people's potential for the future development of the country, their present needs and problems are often disregarded and delegated to their families, as discussed above. The framing of youth as nothing more than a period of transition into adulthood thus points to a pragmatic relationship with young people as citizens. While they are valued for their potential for society, they are invisible as citizens. Or, if they are visible, it is because they pose a threat to social order. An interviewee feels that the state does not see young people until they become a "problem," i.e. take to the streets.

> I only have a name and number in the records, the state follows me around to see if I will go out on the street with a rock in my hand. Otherwise, it does not relate to me, either through health care or education or upholding my right to live. Maybe you will

> be able to make your own way in the future, you either put your life on the right track or you abandon the idea of making your own way, and youth is exactly this stage in the middle. (27, working, leftist and activist in right to the city movements)

Even if young people are continuously pushed to make choices and work for their future careers, available options seem rather limited. Because they cannot obtain sufficient financial support from institutions and rely on their families instead, their future career choices are to a great extent shaped by their families' socio-economic status. For those who cannot count on support from their families, choices of future profession are even more restricted. The following account by a participant who comes from a working-class family shows how young people's paths to the world of work are standardized in a way, how alternative choices are not only limited but expensive, and therefore, very much related to social inequalities.

> You have to go to university, you should be a student, then you need to find a job and keep it. Otherwise, working without a university degree means just working. Or you should be very lucky, or really strive in order to find a way for yourself. For example, if I wanted to be a theater actor, I would not be able to, because I would need to take an exam, or I would get into courses by paying large amounts of fees... And it is really difficult. Everything is about money. You work even while you are a student because you constantly have to pay for something. (23, working, independent activist)

The limited number of educational options is not the only problem, it is compounded with a lack of opportunities in the job market. Even those young people who were able to choose their educational paths frequently end up working in totally different sectors. A participant emphasizes that having a university degree in a certain field does not guarantee a job in that field. Many young people work in jobs they were not trained for, just because there are no other available options that offer a reasonable income and a decent lifestyle. An activist shares her observations regarding young people's transitions from education to work in Turkey:

> If you live in this country, for example, if you have a philosophy degree you almost never find a job related to this, because you have to do other jobs to make money. If you are a doctor or a lawyer it is a bit different, but there are large numbers of young people who cannot work in their fields yet still have to make a living... (27, working, leftist and activist in right to the city movements)

The last two accounts underline the role of socio-economic factors, family background specifically, in setting young people's paths from education to work in

Turkey. The chances and opportunities for young individuals are to a great extent determined by their families' socio-economic circumstances. Below, an interviewee explains how social inequalities shape youth experiences. She acknowledges the role of her families' continuous and substantial support in enabling her to get a good education and eventually find a good job. She is aware that family support is not something all young people can count on, and that there are even some who have the responsibility of contributing to the family budget.

> If you have money to support yourself financially or even something more than just supporting yourself, then being young in Turkey is easy and joyful. However, many young people are trying to support their families, there are many whose families are not well-off, so they have many responsibilities. When you look at them, you see it is not easy. But if you look at me, and young lives like mine, being young is really easy. For example, regarding working life, I easily found a job, because I had a good education, because my family had the money to send me to a private university. Although I was not that successful, I was able to attend this university with the support of my family, then I graduated and easily found a job. That is to say, everything is connected—you should either be exceptionally gifted to enter a top university or you should have money. If you have none of these, it is going really difficult for you. (25, employed, no political affiliation)

Because Turkey's social policy regime leaves the task of supporting young people to the families, family background becomes a critically important factor determining young people's outcomes in adult life when they gain their social autonomy, enter work, and start their own families. As this interviewee explains, for those lacking family support, the transition from education to work, and therefore the whole experience youth, becomes rather difficult.

These reflections show that it is common for young people to be obsessed with the future, and although most of them are critical toward this overemphasis on the future and its effects on their lives, there is still the fact that anxiety about the future is one of the main factors shaping young people's lives in Turkey. One of the interviewees underlines the invalidity of the assumed direct relationship between having a university degree and having a job in Turkey:

> I think that the main point about youth is the anxiety about the future. In Turkey, there are too many unemployed people with university degrees... Students as early as the second year in university start to think about what to do after graduation... After years of education, people cannot take part in the production process. (23, student, nationalist youth organization)

For those who succeed in finding jobs, being young emerges as an important factor determining their position as an employee. Since youth is mostly thought to be synonymous with inexperience and insufficient skill, young people find themselves on the lower rungs of workplace of hierarchies. One participant shares her experience of being made to censor her social media posts about the Gezi protests by her employer. She thinks that what enables the employer to exert this type of pressure on her is the fact that she is a young person and has limited opportunities in the labor market.

> Being young is like this: If I had been a senior figure in that firm, it would not have been the same, I could have a say in the matter. However, I am young, inexperienced, and low-cost labor, therefore I have to put up with things like censorship... They know that I need this job, I have to work, and there are millions of unemployed people like me. (23, employed, neighborhood solidarity)

Another interviewee, a student who works in part-time jobs to make a living, recounts how young people face poor working conditions and are deprived of their rights, especially in the temporary staff system in Turkey. This participant came to Istanbul to attend university. His family is not well-off and he cannot receive any financial support from them. As a student struggling to continue living in Istanbul, he worked on several construction sites and in food service. His experience illustrates how young people fare in the current job market dynamics. Furthermore, his account sheds light on how structural conditions and processes such as migration determine the experience of being young. What is depicted here is not a transition from education to work, but rather being stuck between two identities, that of a student and that of a worker. In order to be able to finance their studies that would benefit them in the future, students enter the job market as low-cost and often exploited workforce. Missing state support, especially if compounded with a lack of support from the family, makes young people one of the most vulnerable groups in the labor market.

> Being young is difficult here. This generation is exploited to a great extent. Talking about myself, I have worked in construction without being paid for weeks. I went to collect my salary, but I was sent back as if I had not earned it. Youth is the period in which instances of oppression and violation are very common. Young people are exploited as low-cost labor... I worked in a restaurant for 8–9 hours (per day) last summer. Before that, I worked at a construction company. And before that, I washed the dishes at a construction site. Sometimes there are jobs like waiting tables in cocktail bars. But I no longer want to do these types of things; I do not want to let myself be exploited, especially in this temporary staff system. There are agencies, they gather people and give them jobs, and then they take exactly half of the money we earn for

themselves, although they do not do the actual work. I will not let myself be used in this way even if I have to go hungry. (22, student, environmental activist)

### 6.2.3 Being Young in Turkey vs. Being Young in Europe

The interviewees in this study often compared European countries with Turkey as dissimilar contexts in which being young is experienced. While the European context is mostly described as offering its inhabitants contentment and peace of mind, Turkey is associated with hardships, suffering, and oppression. Beyond the apparent difference in the levels of youth well-being between European states and Turkey,[5] the interviewees' comparisons paint an idealized picture of young people's lives in Europe. Looking closer at this idealization reveals what the ideal conditions of being young look like for the interviewees, as well as how they perceive the existing conditions of being young in Turkey. One of them gives an account of her Erasmus stay in the Netherlands, emphasizing the peaceful atmosphere in the country. She also highlights how confident young people were in their life choices there. The interviewee's emphasis on young peoples' self-assured and carefree attitudes about educational and occupational choices is contrasted with young people's transitions from education to work in Turkey as a stressful and difficult period in which opportunities are limited and socio-economic background is paramount.

It was such a quiet and peaceful life in the Netherlands, it was sometimes even a suffocating calmness... People were so healthy and happy, and they seemed so determined. Dutch people I met, especially young people, had cleared their own path in life and were going down that path. (23, student, no political affiliation)

Similarly, another interviewee recalls her visit to Germany during her high school years. She underlines the relaxed mood of young people in Germany and their empathetic behavior towards each other. In contrast, being young in Turkey is connected with pressures and worries even in one's leisure time, inability to

---

[5] The Global Youth Well-Being Index, an initiative of the International Youth Foundation, ranks countries according to their scores in areas such as gender equality, economic opportunity, education, health, safety and security, citizen participation, and information and communication technology. According to the 2017 index, Turkey differs remarkably from its European counterparts in terms of youth well-being. For example, while Germany ranks fourth, Turkey ranks twentieth out of 29 countries. What is striking is that Turkey is reported to have the most stressed young people of any indexed country. For more details, see: http://www.youthindex.org.

express oneself, and concerns about one's position in society. The interviewee reports:

> We went to Nürnberg, Germany, for a language course. There was a different youth there. People were incredibly relaxed and were not interfered with each other... They treated each other with much more empathy... Growing up in Turkey, you always feel pressure on you to signal status, show that you are a stylish person, a good person, that you go to the right places, etc. (24, student, environmental activist)

What these young people observed and found fascinating on their travels in Europe points to what they feel is missing in their own country. Calm atmosphere and streets full of healthy, happy, relaxed people are not something familiar to young people in Turkey, and they see these two contexts in sharp contrast with each other. The perception of European youth as happy and confident and living in conditions of comfort and affluence stereotypically assumes that young people in Europe are free from inequalities and hardships linked with both global processes and national-level dynamics. However, it is still explanatory in terms of how the interviewees' see their own youth. One describes being young in Turkey as a position in between Europe and the Middle East:

> We seem like a mixture of European youth who live comfortably with access to technology and welfare and the Middle Eastern youth who experience all sorts of difficulties and misery. (22, student, environmental activist)

Comparing their experiences with those they perceive as faring better than themselves, the interviewees often underlined that being young in Turkey was shaped by unequal relations of power. One of the participants underlines the importance of context in defining the experience of youth, and notes that relations of oppression are indeed one of the most important contextual dynamics of being young in Turkey:

> Children, youth, adolescents... I cannot talk about any age group without contextualizing them within the circumstances in society... When I say youth in Turkey, I cannot think of a specific thing. I just can think of some people spending their lives within relations of oppression. (29, employed, feminist activist)

In terms of oppression, they emphasized the obstacles faced by youth in urban spaces. In the account below, one study participant cites the example of İstiklal Street that ends in Taksim Square and is close to Gezi Park. For a long time, the street had been home to entertainment and arts scene as well as youth cultures and political groups. As discussed in Chap. 4, İstiklal is also one of the places where urban regeneration plans garnered public opposition as they caused the street to lose its meaning and function as a hub of cultural and political events in recent years and put the area under the strict control of the authorities. An interviewee relates that young people no longer feel free in İstiklal and the places they can gather are becoming more scarce by the day:

> When we hangout in İstiklal, police officers come and tell us to move along, make frequent and arbitrary ID checks... They continuously oppress youth, limit access to the public spaces, and control people, and this will eventually lead to an outburst. If a young man does graffiti, but the municipality does not allow this, then this young man will gets angry. (23, part-time employed, independent activist)

The interviewees also occasionally mentioned "going away," in the sense of migrating from the country. As they imagine Europe as a much better place in terms of living standard, working conditions, and everyday life relations, while at the same time they experience the hardships and pressures of the Turkish context, some young people feel inclined towards building a future for themselves abroad. This inclination is undoubtedly accompanied by misgivings, since the decision to leave one's homeland is never an easy one. One of the interviewees shares her reflections regarding the idea of "going away," and underlines how her belief in change prevents her from immediately migrating to some other place:

> It is really hard to be young in Turkey... We constantly think about going away, that is always on our mind. However, if we do not struggle, who will change this country, how will this country change, how will we do this? If we accept every demand they make then no one will be able to live here. Therefore we do not go anywhere even if we think about it. (20, student, feminist and Kurdish activist)

## 6.3 Being Young and Being Political: Mission, and Power

This section focuses on how being young relates to being political. What is the relevance of youth for political action? How is being political discussed in relation to the definitions of being young? What are the implications of being young in political organizations? Guided by these questions, the section analyzes and discusses young interviewees' accounts of being young and being political, in order to understand the way youth is regarded as an endowment of physical and mental capacity and potential for political action. It goes on to take a look at how being young shaped participants' experiences of political action, particularly in political organizations.

### 6.3.1 Potential for Transformation and Political Action

The interviewees often referred to youth as a period in life that brings with it the energy and free time needed for political activism, and described in relation to potential for critical thinking, openness to new ideas, and courage for change. An environmental activist makes a comparison between himself and his mother, and points out how being young makes a difference in terms of one's interpretation of the same socio-political context. The interviewee's mother is from the generation that had experienced the military coups and violence on the streets. Thus, as the interviewee explains, this previous generation might find today's political atmosphere comparatively better than the previous ones, since they see less violent events on the streets and less oppression than in a military coup, and therefore they choose to comply with the current situation. However, since young people do not evaluate the situation relatively, they are "sufficiently" disturbed to criticize the existing political options, and have the belief and will to change something. The interviewee states that:

> For us, I mean for people of my age, there is a chance. I can change my mind, but I cannot change my mother's. Because my mother is happy when she compares today and yesterday, she is really ok with the way the country is governed today. But I am not happy with it, because I believe that it could be better. As a young person, I have not seen anything worse than today, thus I want to change today. (22, student, environmental activist)

Similarly, a feminist and Kurdish activist shares how for her, being young connotes spearheading change, and she invokes the "revolutionary youth" stereotype. She suggests that this discourse on the revolutionary potential of youth is produced by young people themselves, and reflects on how it shapes their approach to politics:

> When I say young, I mean someone who is leading, who should be leading. Someone who is up-to-date, things have changed, you have to keep learning about what you should be struggling against. An older person may think that it is too late, but you are a young person and you believe that you can change something. (22, student, feminist and Kurdish activist)

Another interviewee points out that young people are valued by society and seen as having potential for political action, but their political potential is framed in a pragmatic manner and their political agency is thought to be malleable:

> Being young is glorified in society, because young people can learn, can change their minds; they can be shaped to a great extent... It means that they are a group that can be politicized more easily. (20, preparing for the university entrance exams, leftist and Armenian activist)

Recalling her involvement in the political activities following the Gezi protests when people started to come together in park forums for discussions and political action, an activist underscores the physical advantages that young people tend to have when it comes to political action. Being young means having the necessary endurance and vigor to attend back-to-back meetings and follow many issues at the same time:

> There are some advantages to being young, you can take care of five different things and still have energy left, you can get involved in many things at the same time. For example, I was attending lots of working groups at once. I could go to another meeting after a forum workshop, etc. You can do many more things compared to a fifty-year-old person. (25, employed, leftist and Alevi activist)

### 6.3.2 Being Young in Social Movements/Political Organizations

Young people's accounts also reveal that being political is frequently framed as inconsistent with things universally associated with being young such as having a lot of free time, few responsibilities, or a penchant for having fun and doing crazy things. One of the activists reported how her received political identity shaped her youth:

> When I say youth, I think of going to meetings, expressing your ideas. For example, I did not see this period as a crazy time, such as hitting and breaking. What I have been doing in my youth is going to protests. Actually, not only in my youth, I have done that my whole life... I have been alive for 20 years and I have been going to Newroz[6] for 20 years. I was born into a politicized environment. (20, student, feminist and Kurdish activist)

[6] A spring festival of special cultural and political significance for Kurdish people.

Similarly, an LGBT activist recalls how she started going to protests in her early adolescence and has been politically active ever since. Her youth is "not all about fun and enthusiasm":

> I have been protesting on the streets since I was 13 or 14 years old. I was in the "No to War in Iraq" protests and in the protests against nuclear plants. My organizational experience is quite substantial, and therefore I did not necessarily experience those times as my youth. It was not all about fun and enthusiasm. (23, student, LGBT activist)

These accounts pointing to an incongruity between political engagement and a youthful lifestyle as it is universally imagined and also seem to suggest a division among young people, or rather, different ways of experiencing youth. While politically engaged individuals live through their youth in a more "serious" manner, engaging with social issues and devoting their time to political action, members of the other group live "lighter" lives—that is, they spend time enjoying themselves and traveling and do not busy themselves with social issues. On the one hand, this highlighted contrast between political action and youthful practices reveals that politics and political activity are imagined as phenomena related to adulthood rather than youth. On the other hand, it highlights the limitations of universal assumptions about the concept of youth. The fact that politically engaged young people do not think that they experiencing youth as it is usually experienced is a reminder of the importance of analytical discussions on the context-contingent diversity of experiences of youth that go beyond categorical definitions. One study participant, an activist in the right to the city movement, expresses how he feels distant from the insouciance of youth because he is overwhelmed by his activities in oppositional politics:

> I do not see myself as young, I feel old and exhausted. I need some rest. However, the (oppositional) movement needs people like me, there are lots of gaps, young people are not politicized, young people are apolitical. That is why I am taking part in the struggles in the field of education and urban struggles. I wish giving up were an option. I wish other people would contribute to the opposition instead of taking a neutral position without reflecting on their own ideologies, then I could have a chance to quit and tend to other things. I have not been able to live the experience of being young very much, and I still cannot. Participating in meetings 5–6 times a week, this is not something a young person does. I do not have this experience of being care-free. I am not that unhappy, but I feel tired. (23, student and employed, right to the city activist)

Young people find the burden of political action incompatible with the common conceptions of youth. Beyond that, this account reveals that the perception of young people as having inherent political potential and the mission to be the vanguard of change may place a heavy burden on the politically engaged among them. This interviewee feels that he needs a break, but his sense of responsibility prevents him from relenting in the midst of struggle. Another, an environmental activist, gives a strikingly similar account of her experience of political engagement and being young:

> Being young is the new being old I think... I have started to feel a bit tired recently. People always tell me that being young means being dynamic, having lots of energy, but it requires a big responsibility... The man [a farmer her environmental organization hosted for a talk] is saying that you will do politics, you will do the production—you will make this country a more powerful one. At first you are happy to hear this, but then you feel the weight of the responsibility.... To some extent I see being young as a responsibility. (24, student, environmental activist)

Being young undoubtedly has implications for young people's status in political organizations, as the following account illustrates. While talking about intergenerational conflicts in her organization, a feminist activist criticizes older generations' expectations and approach to young members of the group. Her account draws attention to the relationship between the social conditions of being young and young people's political engagement. Living in a context characterized by insecurity and uncertainty, political practices of the younger generation differ from the politics of previous generations, as might be expected. The interviewee states that:

> You are expected to be active all the time and continue struggling tirelessly, although you have your own life and your own concerns as well. That is to say, as a young person who knows so little about what will happen in your life, faced with so much insecurity, demands for flexibility, potential complications, etc., those people expect you to be energetic. When they were doing politics in their youth it was totally different from the current situation... (29, employed, feminist activist)

## 6.4 Exceptional Modes of Being Young in the Gezi Protests

This final section of the chapter will take closer look at how the notion of youth was negotiated in the context of the Gezi protests. It starts with an analysis of the interviewees' reflections on age-based relationships, particularly their relationships with previous generations in light of experiences of the Gezi protests, and goes on to discuss how young people's perceptions of being young were reshaped by their participation. This sub-section outlines the way the Gezi protests, in addition to bringing about a sense of generational consciousness and recognition by the older generations, seem to have provided a space for new modes of experiencing youth and transformed young people's definitions of it.

### 6.4.1 Generational Consciousness and Recognition

The participants in this study often mentioned that the Gezi protests enabled them to relate with their peers. Whereas before the protests they had mostly led isolated individual lives, the protests made them aware of the fact that they had common experiences and problems as young people who had come of age in recent decades. A participant underlines the value and importance of this generational consciousness:

> What I witnessed was that youth, at least our generation, came together and recognized that we had similar expectations, aspirations, grievances, and concerns about the future. This was so important for us. (28, employed, no political affiliation)

As the interviewee describes, the generational consciousness that sprang up with the protests is built on a perception of shared feelings, thoughts, and practices that came about under the same set of specific socio-political conditions. It brought with it a realization of the social factors that young people in Turkey had been exposed to collectively. Their solidarity-based relationships built during the protests and especially their experiences from the camping period[7] were the main points around which this generational consciousness was centered. Another crucial source of connection were pop culture and humorous and creative online content. References to online games, drama series, or online platforms, all popular with young people, helped the protesters discover their commonalities and similarities of experience and develop a sense of belonging to the protest environment. References to these shared experiences and popular and online culture served to reveal the "passive networks" between young individuals. Described as the "instantaneous communication among atomized individuals that is established by a tacit recognition of their common identity, and which is mediated through real and virtual space," passive networks point to the potential of atomized individuals to mobilize and act collectively without organizations and deliberate networking (Bayat 2013, p. 53). It was precisely in this way, through communication via

[7] Young people's experiences of the protests were discussed in detail in Sect. 4.3.

graffiti, chants, performances, and social media posts and messages in the language and style of the 90s generation that the protesters at Gezi became aware of the unseen networks between them and their potential for collective action. One environmental activist recalls how spending time on the 9gag.com[8] during the protests made him realize he was not alone and isolated:

> People were so happy in the Gezi protests… They became aware that they were not alone, there were many others killing time on 9gag... I was also relieved to find out that I was not crazy! (22, student, environmental activist)

Young people report that the Gezi protests had an effect on their relations with adults around them, and that they lead to renegotiation of the intergenerational conflicts. One activist suggests that his generation of young people was brought up to attach primary importance to their individual concerns and career goals, which explains their avoidance of collective action and social issues. Underlining that this kind of upbringing created pressure for young people, the interviewee interprets the Gezi protests as young people's rebellion against their families:

> These parents raised their children to be too individualistic and career-oriented, I know that many people feel oppressed by this... That is why I think my generation is under severe pressure. What happened in Gezi was not only directed against the state, but also against the family. Our generation had a really hard time coming of age, but Gezi made us grow up. (22, student, environmental activist)

Thus Gezi gave young people who struggled under the burden of their parents' expectations had the chance to express themselves and their own aspirations free from adult interference. Their solidarity-based collective action and their youthful yet effective mode of doing politics also earned them a measure of respect in the eyes of adults. One participant recounts how older people's demeanor toward young people changed during the protests:

> In that period no one tried to give advice to young people. Normally, it is so common to suggest that young people should do this or that. Young people's perception was similar, they (previous generations) could not do anything in their day and now we were going to do better... During the protests it was good. Everyone respected young people. For example, if my sister does not come home at night, it is usually a disaster at our home. However, it was not like that during the protests. When she said that she was in the park, there was no more discussion... It is said that young people found

[8] "9gag: Go fun the world" is an online platform for hosting and sharing memes, humorous photos, and videos. For more, see: https://9gag.com.

> the chance to express themselves. Gezi was an important period for young people's self-expression and self-confidence. (25, employed, leftist and Alevi activist)

Undoubtedly, in some cases young people's political action in the protests was not so influential as to change the previous generations' perceptions of the young generation. The interviewees recounted that the adult world around them, particularly their parents, frequently dismissed young people's political agency and underestimated the potential of their new approaches to politics and collective action. An interviewee remembers her father's disparaging comments about the Gezi protests. Her father is from the 70s generation that experienced turmoil at universities and on the streets, and in making comparisons with that period, he tended to downplay the potential and importance of young people's political action in Gezi.

> Yes, I felt incredibly good there (in the Gezi Park), I felt confident, but then it is different in the world of adults. They always say: "we have seen so many things," I do not know exactly which important events they witnessed, we are reading history, too, but their historical events are much more valuable to them since they experienced them firsthand. So, I felt that I was hitting the wall of reality, especially in my discussions with my father. (23, student, no political affiliation)

At the same time, there are examples of young people's political potential being acknowledged, but this acknowledgment came with expectations attached and placed the burden of political change on young people. The new generation has been seen as lucky because Gezi gave them an opportunity to participate in a major political event and practice collective action on a large scale, and in the period that followed they were expected to make use of this political experience and carry on being politically active. As discussed in Chap. 4, many young people were disappointed and felt hopeless after the protests, and their disappointment seems to be related to the burden of expectations they carried, which, in conjunction with the intense efforts invested in political action, made them feel that they were being robbed of their youth:

> For me, being young is a great thing. But I think we were not allowed to be young, especially after Gezi. Expectations were so high, especially of the 90s generation. For example, our friends aged 30 and over coming from organized political struggle, they saw us as lucky. They thought we were lucky to have experienced Gezi. (23, employed, right to the city activist)

### 6.4.2 Transformation of the Perception and Experience of Youth

In addition to fostering a generational consciousness among young people and changes in intergenerational relations, the Gezi protests were also a space in which young people revised their notions of being young and discovered new ways of experiencing youth. Unsurprisingly, the protests led many of them to realize their own political potential. As a generation whose political agency had been played down and ignored, they gained the confidence to air their grievances and demands and expect success in achieving their goals. A study participant underlines how the protests and political activities in the local initiatives enabled young people to express themselves. She also emphasizes the importance of the sense of belonging, a feeling of attachment and participation in the collective identity imagined in the protests. Her account is informed by her experience as a volunteer in a civil society organization working on youth.

> Young people who previously lacked this confidence found it there [in the Gezi protests]. It is such an important thing to feel that you belong to something. Everything else aside, there was a sense of belonging there. And young people were capable of expressing themselves, unlike before... They had been feeling worthless. (23, employed, youth work volunteer)

Another interviewee shares how she has realized her potential and power as a young person, and how the meaning of being young changed for her after the Gezi protests:

> For me, being young used to be about doing what I want and traveling around. However, it started to mean something else. Youth is full of potential and power, and I think that we should use this to do something. This generation should change something. (24, student, environmental activist)

Study participants also report that their ways of experiencing and performing youth were affected by their participation in the Gezi protests. That is, protests helped young people develop a self-reflexive way of relating to their experiences of being young. The critical potential emerging from the protests seems to have moved them to question how age, young age in particular, shapes their life experiences and relations with others and defines their role in society, subjecting them at the same time to oppression and putting obstacles in their path. These critical reflections were followed by action, small steps toward changing the hierarchical and oppressive relationships in their lives. An environmental activist describes the transformation in her way of experiencing being young by underlining that she no longer feels responsible to meet the expectations of others:

> The protests transformed my perspective about my experience of youth. Now I live for myself. I do not have to explain things to others. I am still aware of the pressures of my environment, but I do not want to cave in to them anymore. I was chatting with a friend a few days ago about the meaning of freedom. For me, freedom is being free from the feeling of oppression. I am struggling for this for myself. I still feel that pressure, but I never felt obliged to react before, I feel much freer compared to previous periods... It is a really fortunate thing that Gezi happened, I am grateful to all these initial protesters who started the protests. (24, student, environmental activist)

As discussed in the previous chapter, the concept of youth and the experience of being young are to a great extent determined by the context in which young individuals live. This chapter has showed that being young in Turkey is characterized mostly by hierarchical relationships, social pressures, anxiety about the future, and inequalities in education and work. Before the protests, these pressures were manifested in authoritarian policies and crackdowns on young people's spaces. In contrast to this, the protests seemed to enable young people to experience an exceptional mode of being young. An LGBT activist recounts:

> Indeed, being young is so hard, but it was easier in Gezi Park. But it was a two-week dream, we woke up, and it was hard again. Many things were easy there, everything was in our hands, we were able to go and eat something, we were able to grab a book from the library in the park and have a look at it, we were able to play volleyball in the park, we were able to dance the *halay* or the *horon*,[9] we were sharing with the others, bonds were strong, but then when it finished, we returned to our usual lives, to our lives in Turkey. (20, student, LGBT activist)

As the interviewee points out, meeting their basic needs via the solidarity infrastructure, having the chance to take a book from the library for free, dance, and socialize with many others were all different aspects of the unique experience young people had in the protests. Coupled with the practices of self-expression and political activity, the Gezi protests transformed youth into a vigorous experience. Instead of a youth defined by hardship, obstacles, and oppression, new ways of being young characterized by freedom, solidarity, and change became possible. Portraying the way youth was lived in the Gezi protests as a "dream," the interviewee emphasizes that this exceptional mode of being young was an

---

[9] *Halay* is the traditional folk dance in the east of Turkey; *Horon* is the traditional folk dance of Turkey's Black Sea region.

illustration of young people's aspirations, desires, and what they imagined and longed for. The way in which the Gezi protests affected the participants' experiences of being young enables an interpretation with reference to "youthfulness" as Asef Bayat conceptualizes it in his discussion on young people and social movements in the Middle East (Bayat 2013). Bayat argues that "a youth movement is about reclaiming youthfulness" (Bayat 2013, p. 17), and explains how youthfulness becomes the defining characteristic of a movement:

> It embodies a collective challenge whose central goal consists of defending and extending youth habitus—defending and extending the conditions that allow the young to assert their individuality, creativity, and lightness and free them from anxiety over the prospect of their future (Bayat 2013, p. 18)

In light of Bayat's discussion, the concept of youthfulness is useful in explaining the context of the Gezi protests. The analysis in this chapter of first-hand accounts strongly indicates that Gezi provided a "youthful space" for the protesters. The argument is not that Gezi as a youth movement was about reclaiming youthfulness. Rather, the protests were an alternative temporality and spatiality in which young people discovered their power and capacity as individuals, their potential for collective action, and the effectiveness of their humorous and creative use of language and youth culture. Experimenting in a more supportive and sympathetic atmosphere gave them necessary respite from the anxieties, pressures, and hardships of their everyday lives as young people in Turkey.

## 6.5 Conclusion

This chapter has examined young people's reflections on being young. Relying on the sociological and anthropological approaches to studying youth, the chapter investigates it as a relational and context-based concept. The participants' experiences of being young are studied with regard to their circumstances in life as well as in relation to their social environments, and special care is taken in the analysis of the field data to account for the structural factors that shaped young individuals' experiences at Gezi Park.

The chapter begins with young people's accounts underlining the positive potential and capabilities inherent to being young. This section reveals that the universal characteristics attributed to youth, such as high energy levels, dynamic lifestyle, freedom from responsibilities, etc. are limited by the structural conditions in which one lives, and are therefore far from being innate, natural characteristics of young people in general. Despite references to the potentials and capabilities that youth brings, the interviewees mostly describe being young in Turkey in relation to hardships, restrictions, and hierarchies. Thus, the chapter's second section focuses on these "negative" accounts of being young. First, it discusses how age mediates young people's everyday lives through intergenerational relations, social pressure, and gender. Second, it addresses young people's accounts of their common experiences of anxiety about the future, and shows that the participants' choices, actions, and decisions in their youth are shaped by their concerns about securing a future for themselves, which usually meant a good job. This is followed by accounts of the transition from education to work where the interviewees report feeling fettered by the limitations of their families' socio-economic circumstances in the absence of social policies supporting young people, as well as discussion focused on comparisons between the conditions of being young in Turkey with those in Europe. The third section explores the interviewees' perceptions of being young in relation to being political. They see being young and having physical and mental potential for political action as strongly related, which indicates that they seem to internalize and naturalize the widely held assumptions about the political and revolutionary potential of youth. The discussion then examines the interviewees' narratives of being young in social movements and political organizations, and looks more closely at how age-related relationships shape collective and political action. The last part of the chapter focuses on the impact of Gezi on young people's perception and experience of being young, in light of the generational consciousness they forged and the respect from the older generations they earned in the protests. They report dramatic change in their experience of youth that happened as soon as the limitations and restrictions from their everyday lives no longer applied.

This chapter focuses on the relation between experiencing youth as a social relationship and young people's engagements in political action and membership in social movements. Their perceptions and experiences of being young in Turkey prompted them to join the Gezi protests, and the transformative effect of the protests in turn enabled them to experience youth in new ways and change their age-based relationships with adults. With this in mind, the value of analyzing young people in the context of the Gezi protests lies in discovering how being young impacted their involvement and ways of acting in the protests, rather than gauging whether or not their participation in the protests points to a massive re-politicization of youth in Turkey or the birth of a revolutionary generation. This avoids treating youth as a mere age-based demographic category of a given social movement; rather, it must be seen as a social category that is dynamically and continuously constructed in the course of collective action. Therefore, studying how being young was transformed in the protests provides insights not only into young people's experiences, but also into the ways in which they relate to the context they live in, and the dynamics which set the background for their participation in a movement. This seems to be the crucial point at which the intersection of the fields of the sociology of youth and the study of social movements promises new perspectives for understanding young people, collective action, and society in general.

## References


Abbott-Chapman, J., & Robinson, M. (2009). Leisure activities, place and identity. In A. Furlong (Ed.), *Handbook of youth and young adulthood: New perspectives and agendas* (1st ed.). London and New York: Routledge.

Alemdaroğlu, A. (2005). Bir İmkân Olarak Gençlik. *Birikim, 196,* 21–30.

Bayat, A. (2013). *Life as politics: How ordinary people change the Middle East*. Stanford: Stanford University Press.

Beck, U., & Beck-Gernsheim, E. (2002). *Individualization: Institutionalized individualism and its social and political consequences*. London, Thousand Oaks, New Delhi: Sage Publications.

Bourdieu, P. (1993). Youth is Just a Word, In *Sociology in question* (R. Nice, Trans.) (pp. 94–101). London, Thousand Oaks, New Delhi: Sage Publications. (Original work published 1984).

Candan, C. (2008). 3 Hours [3 Saat: Bir ÖSS Belgeseli], video recording. Giyotin Film.

Çelik, K., & Lüküslü, D. (2010). Spotlighting a silent category of young females: The life experiences of 'House Girls' in Turkey. *Youth and Society, 44*(1), 28–48.

Furlong, A. (2013). *Youth studies: An introduction.* London and New York: Routledge.

Lüküslü, D. (2005). Constructors and constructed: Youth as a political actor in modernising Turkey. In J. Forbrig (Ed.), *Revisiting youth political participation* (pp. 29–36). Strasbourg: Council of Europe Publishing.

Yılmaz, V. (2016). Youth welfare policy in Turkey in comparative perspective: A case of 'Denied Youth Citizenship.' *Southeast European and Black Sea Studies, 17*(1), 41–55.

Yurttagüler, L. (2014). Gençlerin Özerkliği Var Mı? In L. Yurttagüler, B. Oy, & Y. Kurtaran (Eds.), *Özerklik ve Özgürlükler Açısından Türkiye'de Gençlik Politikaları* (pp. 11–36). Istanbul: Bilgi University Press.

# Part III
# Politics of Youth

# Studying Youth and Politics

# 7

This chapter examines how young people's political participation has been studied in the literature as well as how the concept of the political has been defined in recent theoretical discussions. It explores these topics in accordance with this study's overall approach, and makes an overview that constitutes the theoretical background to the next chapter that discusses young people's politics that emerged from the Gezi protests. The two chapters offer an in-depth analysis of youth and politics in Turkey based on the accounts of the young participants of the Gezi protests along with an exhaustive examination of their concept of the political.

The first part of the chapter investigates how literature on political participation has been criticized, especially with respect to its conceptualizations of the definitions of participation and politics and its construction of the binary opposition of political and apolitical young people. By navigating these critical discussions, this chapter explores new perspectives for discovering young people's complex and diverse ways of relating to politics, and takes a look at the ways in which young people's relationship with politics in Turkey has been studied.

The second section focuses on the notion of the political. In line with the critical viewpoints that question traditional approaches that tend to restrict the act of doing politics and the notion of the political to specific forms and places, this section emphasizes the new modes and spaces of politics with reference to recent theoretical discussions about the definition of the political. In addition, it scrutinizes the way the politics of the Gezi protests have been interpreted in the light of the recent discussions, particularly from radical democratic theoretical positions.



P. Gümüş Mantu, *Redefining the Political: Youth Experiences of Collective Action in Turkey*, Kulturelle Figurationen: Artefakte, Praktiken, Fiktionen,
https://doi.org/10.1007/978-3-658-40565-6_7

## 7.1 Youth Political Participation: Beyond Binary Oppositions

Young people's relationship with politics has long been discussed primarily through the concept of participation, and their non-participation has been problematized on the global scale. Low levels of young people's political participation, and their lack of interest in conventional political processes, have been at issue in many studies. When it comes to politics, it is usually pointed out that young people's political apathy, their lack of awareness and participation in the political process, poses a threat to democratic traditions (Furlong and Cartmel 2007, pp. 121–122). Moreover, young people's disinterest in politics, whether described as apolitical, apathetic, or disengaged, has been framed as a generational problem in the popular discourse. The generation that grew up in the post-Cold War period, in the 1990s and 2000s, has generally been perceived as individualistic, competitive, consumption-oriented, uninterested in collective solutions to problems in society, and ultimately uninterested in political participation, particularly due to being subjected to the ubiquitous effects of globalization and extensive use of communication technologies.

This mainstream way of understanding young people's relationship with politics that focuses on young people's lack of participation as well as the definition of participation itself, has been widely questioned and criticized. Judith Bessant draws attention to structural barriers—such as social and economic injustice or a lack of freedom of movement, speech, and assembly—that hinder young people's political participation (Bessant 2004). Bessant reminds us that youth political participation cannot be decontextualized, since it is always intertwined with structural conditions. Therefore, a discussion on participation that does not take social and economic conditions and basic rights into account is not only limited but also useless.

The study of changing trends in young people's relation to politics has also contributed to the critical discussions in the field of political participation. On the basis of a fifteen-nation European Social Survey, Pippa Norris writes that younger generations prefer to participate in cause-oriented repertories such as consumer politics, demonstrations, and petitions, rather than citizen-oriented repertories such as voting, party work, and contact activity. Moreover, there has been a transition from traditional voluntary associations such as churches, unions, and political parties towards new social movements and advocacy networks, including environmental and humanitarian organizations (Norris 2004, p. 22, Fig. 1). Similarly, Lasse Siurala describes a trend towards postmodern or emerging and future forms of participation, mostly characterized as emotional, expressive, and esthetic forms of engagement. He discusses the socio-economic basis of this trend, identifying three main concepts: insecurity, changeability, and transience (the latter two in reference to Melucci 1996). Young people do not feel secure about their future; however, they have an increased awareness of the possibilities around them and they feel that they can experiment. They therefore concentrate on experimenting with their individual capacities—the most important thing for young people is the present (Siurala 2000). Cristina Fomiyana underlines that a narrow definition of political participation, limited to elections and formal political institutions, creates the false impression of low participation or interest in politics on the part of young people. However, when participation in civic and voluntary associations and in protest and social movements is considered, the picture changes (Fominaya 2012).

Although many of the survey-based quantitative studies show that youth political participation is in decline, or at best limited, most of them fail to explain the reasons for this withdrawal. Qualitative studies, on the other hand, go in detail in terms of explaining why and how young people have become distant from formal politics. Nevertheless, the literature in this field generally tends to define politics and the political using a top-down approach, therefore equating the non-participation in those arenas to political apathy, without even asking young people what the political means to them (O'Toole et al. 2003). Rys Farthing engages critically with earlier literature on youth political participation, which tends to rely on binary oppositions such as apolitical/political and engaged/disengaged young people and eventually hinders the development of a complex and thorough perspective on young people and politics. Bringing Ulrich Beck's vision of today's young people as "radically unpolitical" to the discussion on youth political participation, Farthing describes the contemporary phenomenon of young people's disinterest in politics as "the politics of youthful antipolitics." She invites us to see young people's persistent attitude of keeping a critical distance from conventional political processes—even if they are engaged in new forms of political participation—as the very constitutive element of their politics:

> To simply suggest that young people are being political—just in new forms—denies the critical transformation happening in young people's actions. It also overlooks and disempowers young people's *active rejection* of traditional politics. Young people are turning away from politics; they are not just participating in politics in new forms. This rejection needs to be appreciated. (Farthing 2010, p. 188, emphasis in original)

Young people relate to the political sphere in several ways and in different modes. Their "not relating at all," as discussed by Farthing, reveals much about why and how they reject traditional politics, as well as what types of alternative definitions of the political they develop. Defining the political has recently been pointed to as a crucial area of discussion concerning young people's relation to politics. Siyka Kovacheva argues:

> First of all, youth participation research needs to question the concept of the political in the same way as gender researchers have done previously by disputing the established border between the public and the private. (Kovacheva 2005, p. 26)

Thus, similar to questioning the borders between the private and the public, a critical rethinking of the assumed boundaries of the political is meaningful and useful especially for interpreting young people's changing relations with politics. This kind of perspective will not only facilitate deeper understanding of youth and politics, but also contribute to theoretical discussions on the concept of the political.

### 7.1.1 Youth Political Participation in Turkey

In parallel to the international literature, discussions on young people and politics in Turkey also mostly concentrated on the lack of youth political participation. Quantitative studies of young Turkish people's engagement in parliamentary and party politics and political and civil society organizations, mostly show that youth political participation is limited (KONDA 2014; Konrad Adenauer Stiftung 1999).

This lack of youth political participation in Turkey has frequently been explained in reference to the notion of "apolitical youth." The term refers to the generation that grew up after the 1980 coup, when society was depoliticized in all domains and neoliberal policies started to change both economic and social dynamics. In this context, the new urban youth—in other words, "the apolitical youth"—in Turkey has been imagined, similar to their peers worldwide, as a generation for whom money and consumption are crucially important but politics not so much. However, it is worth taking into account that this generation's distancing from Turkish politics has partly been caused by the military coup's oppressive and long-lasting impact on society.[1]

Alternatively, in order to go beyond the insufficiency of evaluations according to which the young are apolitical or apathetic, researchers in the fields of

[1] For a detailed analysis of how the socio-political environment in the post-1980 period affected young people's engagement with the political action and their relationship with politics, see Sect. 3.2.1. Youth Collective Action in Turkey.

sociology and cultural studies have pointed out the need to look more closely at young people's concerns. Demet Lüküslü, on the basis of extensive field research, argued that the post-1980 generation in Turkey is aware of the political arena, and is unhappy with it; members of this generation consciously choose not to participate. Instead, they use coping tactics in everyday life that can be defined as "necessary conformism": humor, making fun of serious issues, or not paying attention at all, all of which allow them to persist in the middle of this uneasy social situation (Lüküslü 2009). Analyzing the public discourse on youth in Turkey, Neyzi states that the post-1980 generation, in contrast to previous generations, is attempting to construct their subject positions by expressing their identities, engaging in cultural practices, and participating in civil society. In that sense, the diverse newly emerging youth cultures in Turkey are growing in parallel to new political identities (Neyzi 2001). Therefore, the popular discourse that judges young people for being apolitical is not only incorrect but also fails to recognize how people are active and motivated in terms of making decisions about their everyday lives and shaping the society they live in, even if this is not truly politics according to the mainstream understanding of the term. In light of Neyzi and Lüküslü's discussions, critical engagement with the concepts of participation and political becomes essential for this study with respect to its investigation of youth and politics in the case of the Gezi protests in Turkey. Young people's participation in the Gezi protests was unexpected and surprising for most of the public, since these young people were members of a generation that had long been regarded as apolitical, disengaged, and even politically incapable in the mainstream discourse. Thus their active presence in the protests delegitimized the apolitical/political and engaged/disengaged dichotomies and revealed the need to understand young people and their relationship with politics in all its complexity, beyond the limits of the existing literature. Good understanding of the meaning of the political and the diverse ways young people relate to politics is essential for this research, as it makes it possible to understand young people's participation in the Gezi protests by taking into account their changing relations with the political.

## 7.2 Defining the Political

As discussed above in relation to youth political participation, how the political is defined is undoubtedly crucial when researching young people and politics. Following Therese O' Toole, Michael Lister, Dave Marsh, Su Jones, and Alex McDonagh's methodological approach to conducting field research with young people with the aim of understanding their conceptions of the political and their political experiences, the term political is used here rather broadly, avoiding narrow, top-down definitions and restricted definitions. Expanding the concept of the political in this way, O'Toole et al. focus on how individuals themselves conceive of and interpret politics as a "lived experience," referring to what they think is political in their lives, rather than responding to adult-made definitions of the political and politics (O'Toole et al. 2003). Along similar lines, this research attempts to grasp the diversity of young people's descriptions of the political and the multiplicity of the ways they see themselves as connected to politics, without definitional limitations. This means that politics and the political are not restricted to any kind of space, activity, or institution. In keeping with young people's definitions of the political, the discussion of politics encompasses perceptions, everyday life, values, and practices. Methodologically, it aims to discover young people's modes of approaching and recreating the meanings of the political and performing politics in all its complexity and multiplicity, without reducing their perceptions and practices to predefined theoretical schema.

Studying the notion of the political opens a wide range of theoretical and political discussions in several disciplines under the umbrella of social sciences. In order to clarify its approach to interpreting young people's definitions of the political, this research focuses on two questions while navigating through the literature: First, where is the political? That is to say, where is the political found? Second, how do we define the political? In reference to which processes, institutions, or practices? The theoretically informed answers to these questions presented here contextualizes young people's relationship with the political by discussing contemporary society and culture.

Globalization has been discussed as one of the most important processes shaping our lives today. It is a complex set of relations whose effects we experience in every aspect of our lives, therefore how we relate to politics is also embedded in the discussion on globalization. Anthony Giddens, in his book *Runaway World*, underlines the contradiction between the way orthodox democracy functions and people's experiences of the globalized world. There is an inconsistency between the way institutions usually function and the way people actually deal with the effects of globalization. People, especially young people, are not unaffected or alienated, but they are concerned with issues such as "ecological questions, human rights, family policy, and sexual freedom," which the current political system has limited means of coping with (Giddens 1999, pp. 92–93). Giddens points out the areas where young people's relation to the political can be analyzed in detail. Moreover, he introduces the notion of "democracy of emotions" as an important characteristic of all types of contemporary social relations.

Democracy of emotions does not mean all partners in a relationship—be it family members, friends, or romantic partners—are always materially equal; however, it is equality in principle. Children in the family are treated as equals and always have the right to respond, in contrast to a state of pure obedience (Giddens 1999, p. 81). Democracy of emotions may explain young people's politics in two ways: first, it underlines the need and search for democracy in different kinds of relations in everyday life. Second, the concept highlights the importance of freedom to act as an equal member of society and express oneself. As will be discussed in the next chapter, young people's insistence and desire for equality in the political sphere, along with their need to express themselves freely in their participation, are distinct features of their relationship with the political revealed by this research.

Ulrich Beck discusses the reinvention of politics in relation to processes of individualization and globalization. In the transition from industrial society to risky modernity, "'individualization' means the disintegration of the certainties of industrial society as well as the compulsion to find and invent new categories for oneself and others without them" (Beck, 1997, p. 95). Beck underlines individualization as a process and struggle of managing new and unexpected risky life situations in which the individual is thought to be the "*actor*, *designer*, *juggler*, and *stage director* of his own biography, identity, social networks, commitments, and convictions" (Ibid, p. 95 emphasis in original). This definition of the individual also requires a rethinking of the political aspect of individuality. As all the institutions of industrial modernity go through a transformation in the process of reflexive modernization, so too does the political sphere. Moreover, similar to other institutions such as the family or economy, the transformation in politics is subtle and difficult to analyze and understand via old categories. Beck writes on the changing meaning of the political:

> In other words, politics breaks open and erupts *beyond* the formal responsibilities and hierarchies, and this is misunderstood particularly by those who unambiguously equate politics with the state, the political system, formal responsibilities and full-time political careers. An ambivalent, multi-level, 'expressionistic concept of politics' (Jürgen Habermas), which permits us to posit the social form and politics as mutually variable, is being introduced here for a very simple reason. It opens a possibility in thought which we increasingly confront today: the political constellation of industrial society is becoming *un*political, while what was unpolitical in industrialism is becoming political. This is a category of transformation of politics with *unchanged* institutions, and with intact power elites that have not been replaced by the new ones. (Beck 1997, p. 99, emphasis in original)

This understanding of the transformation of politics informs the questions of where to look for politics and how to define the political that this study poses. Looking at traditional institutions such as parliaments, political parties, trade unions, etc. in order to understand the change in the political sphere and the meaning of the political is not only narrow, but also invalid and misleading. However, employing the notion of "subpolitics" makes it possible to discover many areas of society where the political is enacted and to explore new ways to define the political that go beyond the traditional conceptions. Beck distinguishes subpolitics from politics in two points: First, professional and occupational groups, the technical intelligentsia in companies, research institutions and management, skilled workers, citizens' initiatives, and the public sphere can participate in the social design as agents outside the political and corporatist system; and second, individuals also take part in the power struggle to shape politics (Beck 1997, p. 103). The emphasis on the individual and subpolitics is crucial for analyzing young people's politics for the present study. Considering young people's individually motivated notions of the political, Beck's understanding of the changing role of the individual in politics enables an alternative understanding of the political in young people's lives without equating politics with collective action and conventional processes. Therefore, young people's individual way of doing politics is still considered politics, and presumably reflects the change in the political sphere. In addition, the concept of subpolitics offers a framework for analyzing young people's definitions of the political with regard to various spheres in which they think they participate in shaping politics. Accordingly, this work uses the notion of the subpolitical to understand the new spaces and genuine ways in which young people engage with the political and to underline the difference between these practices and the conventional forms of politics.

### 7.2.1 The Politics of the Gezi Protests

This final section of the chapter focuses on theoretically informed discussions on the politics of the Gezi protests. It is crucial to note that this section does not claim to define the politics of the protests on the basis of theoretical inquiry; rather, key conceptual references this research relies on in examining young people's definitions of the political are presented here as theoretical background. As discussed above, this study intentionally employs an open approach to the definition of the political, aiming to provide the necessary analytical space for in-depth discovery of young people's perceptions of politics without regard to any preconceived definitions and explanations. Guided by the above discussion of emerging subpolitical spaces and modes of doing politics and the impacts of individualization and globalization, this study pays special attention to how young people as individuals describe their relationships with their social environments as well as how they reflect on their life conditions, and eventually how these relate to their new ways of defining and doing politics. The question that this section seeks to answer against the background of these theoretical foundations can be phrased as: On what terms do the protests provide a lens through which young individuals' relationship with the political can be deciphered? The summary review of theoretical discussions with regards to the politics of the Gezi protests will serve to answer this question.

Making use of Alain Badiou's conceptualization of "event," Spyros Sofos suggests reading Gezi not as a movement but as a moment "in a relatively long sequence of moments, as part of a constellation of contention, or to use the term preferred by Badiou (2005), of *events* that interrupted the otherwise linear time of Turkish politics over the past few years" (Sofos 2017, p. 70, emphasis in original). Sofos points out the political potential of the space and temporality created in the course of the Gezi protests, and simultaneously underlines its broader background of other contentious moments. He describes the Gezi protests as follows:

> Gezi provided a setting, or a happening, a fleeting moment of collective action, a temporary laboratory of meaning creation, of expressing the aporia of its participants during a moment of authoritarian reconfiguration of the institutional architecture of Turkey's political system and economic model (Sofos 2017, p. 70)

Thus, viewed as an event or moment, the Gezi protests, as an unexpected and sudden popular uprising, sparked the imagination and created the possibility of a new politics. This space for possibilities functioned as "a temporary laboratory of meaning creation," as Sofos puts it, and it fostered young people's critical relation to the political motivating them to think of alternatives, even of completely new ideas of the political. Therefore, the Gezi protests opened up a space for young people to discover and experience new ways of relating to politics and the political beyond the limits of the conventional political thinking and practice.

Ayhan Kaya emphasizes the commonality between the Gezi protests and their global counterparts in terms of employing prefigurative politics. He asserts that all these protests refer to an "ideology of change" in search of "another world":

> I argue that it is this kind of prefigurative politics, which incarnates the ideology of these new global social movements experiences in different parts of the world, i.e., *the ideology of change.* The common motto of social protests in Egypt, Turkey, Brazil was "Another World is Possible." The idea of transforming this neoliberal world into

> a world which is more humane, benevolent, compassionate and cohesive is the main source of inspiration of this ideology of change (Kaya 2017, p. 3).

Funda Gençoğlu Onbaşı accentuates that the Gezi protests prompted inquiry into the definition of politics and the political, and points to other, unconventional ways of doing politics and the importance of individual political action (Gençoğlu Onbaşı 2016, pp. 279–280). In her analysis informed by radical democratic theory, Gençoğlu Onbaşı criticizes essentialism, which discusses the political identities in the Gezi protests as complete and fixed. As an alternative, she proposes an understanding of political subjectivities as relational, continuously constituted in the political realm, and open to new articulations (Gençoğlu Onbaşı 2016, pp. 281–282).

Chantal Mouffe, in her analysis of contemporary radical politics, discusses two different political strategies that have recently been influential in social movements: withdrawal from institutions and engagement with institutions. With reference to Michael Hardt and Antonio Negri's political theory of the multitude, "withdrawal from institutions" implies a total rejection of the established political institutions in order to foster autonomous organization of revolutionary forces in society, that is, the multitude. Criticizing this approach, Mouffe suggests engagement with the institutions lined up with her theory of agonistic politics. Mouffe defines the political with reference to its antagonistic potential:

> This is why politics always takes place in a field crisscrossed by antagonisms. To envisage it as 'acting in concert' leads to erasing the ontological dimension of antagonism that I call 'the political', which provides its quasi-transcendental condition of possibility. There will always be a struggle between conflicting hegemonic projects aiming at presenting their views of the common good as the 'true' incarnation of the universal. No rational resolution of the conflict will be ever available. As far as political critique is concerned, it can never be merely oppositional or conceived as desertion because it always engages with a certain aspect of the existing hegemony in order to disarticulate/re-articulate its constitutive elements. (Mouffe 2013, p. 79)

Thus, radical politics cannot and should not aim to end the conflict and arrive at a point of ideal resolution. To be political means to be in a continuous, antagonistic relationship with the hegemony or several hegemonic positions. The present study does not aim to analyze young people's politics in their strategic dimension. Therefore, whether their politics lean toward withdrawal from the institutions or engagement with them is not the central question here. However, Mouffe's approach regarding the antagonistic nature of the political is crucial in thinking young people's relationship with the political. The characteristics of politics emerging from the Gezi protests, as discussed below, indicate that young people imagine political struggle as continuous practice with a basis in everyday life. For the most part, they do not imagine a definitive resolution of the problems, have an antagonistic relationship with the political, and are highly critical of revolutionary perspectives that expect the elimination of problems such as gender inequality, destruction of the environment, or age hierarchy, along with a change in governing bodies. İrem İnceoğlu describes the radical democratic opportunity that emerged in the protests, highlighting the importance of the protesters' engagement with local issues and newly formed local initiatives:

> What we saw at Gezi Park and at the local park forums was exceptional: urban citizens were reclaiming the commons for radical democratic formations while in the process of establishing agonistic relations. These occasions are interesting indicators of a potential radical democratic trajectory: public spaces, in being utilized as forums and public assemblies, offer the potential for creating a space for solidarity that is negotiated around local issues despite ideological, class and other differences. (İnceoğlu 2014, p. 26)

One of the essential features of the political opportunity created by the Gezi protests is that it included premises of antagonistic politics, articulated in anti-hegemonic and critical thought and action. The political potential of the protests, rooted in their prefigurative aspects, was predicated on the belief in the possibility of another world, and focused on the local and the public as sites political action.

## 7.3 Conclusion

This chapter has examined the way young people's relationship with politics has been treated by scholars who study political participation. It has covered discussions on the notion of "the apolitical youth," particularly in Turkey, with reference to approaches that interpret low rates of young people's political participation as a consequence of apathy and lack of interest in political issues. Following the critical perspectives on the conceptualization of political participation and the concept of the political in relevant literature, this study defers to a youth-informed definition of the political. Accordingly, the political as a concept is treated in an open and flexible way, as something to be constructed in the research process in discussion with the interviewees.

The processes of individualization and globalization and the highly individualized perception young people had of their own lives, including their chances, successes, and failures, affects the way they define and reflect on the notion of the political. Thus, their lack of interest in the traditional ways of political action and organization emerges as an implication of broader social change. Similarly, young people's ways of relating to their environment are remarkably different from those of the previous generations. On the one hand, new communication technologies, online spaces, increasing opportunities for mobility and, on the other hand, the side effects of economic globalization, growing inequalities, and deteriorating labor market conditions all shape young lives and mostly create uncertainties, contradictions, and insecurity. All these opportunities and limitations give rise to young people's consciousness and inform their practices as well as their collective action and politics. The global wave of protests in the last decade, including the Gezi protests, is an instance of these circumstances at play in forming young people's political agency, and it is within these social movements that young people's new ways of relating to the political can be traced and analyzed.

The final section of this chapter focuses on discussions on the politics of the Gezi protests. Framing the protests as a space of political possibilities, it stresses the critical potential that transcended the confines of conventional politics and the prefigurative character of the politics of the protests, as well as the local public spaces as centers of theorization, experimentation, and building a better world.

How did young people construct this political space of possibilities? How did their individual demands, grievances, and actions transform into components of the politics of the Gezi protests? How did their relationship with conventional political arena affect the emerging politics of the protests? Taking these initial questions as its point of departure, the next chapter delves into to the analysis of young participants' perceptions of the political and their relationships with politics. It studies young people's political perceptions using a process-based approach, looking more closely at how their relationships with the political were formed/changed/reconfigured throughout the protests, and in doing so it develops a dynamic understanding of the way Gezi transformed youth perceptions of the political.

## References


ARI Düşünce ve Toplumsal Gelişim Derneği. (2001). *Türk Gençliği ve Katılım. [Turkish Youth and Participation]*. International Republican Institute.

Beck, U. (1997). *The reinvention of politics: Rethinking modernity in the global social order* (Mark Ritter, Trans.). Cambridge and Malden: Polity Press. (Original work published 1993)

Bessant, J. (2004). Mixed messages: Youth participation and democratic practice. *Australian Journal of Political Science, 39*(2), 387–404.

Farthing, R. (2010). The politics of youthful antipolitics: Representing the 'issue' of youth participation in politics. *Journal of Youth Studies, 13*(2), 181–195.

Fominaya, C. F. (2012). *Youth participation in contemporary European social movements*. Report for EU-CoE Youth Partnership. Retrieved January 7, 2022, from https://pjp-eu.coe.int/documents/42128013/47261662/Youth_Participation_in_Contemporary_European_Social_Movements_Flesher_Fominaya_FinalPB_CFFPB_CFF_x2x.pdf/30c4dbcd-296e-411f-a338-ac033b3a98f6.

Furlong, A., & Cartmel, F. (2007). *Young people and social change: New Perspectives* (2nd Edition). Berkshire: McGraw-Hill/Open University Press.

Gençoğlu Onbaşı, F. (2016). Gezi park protests in Turkey: From 'Enough is Enough' to counter-hegemony? *Turkish Studies, 17*(2), 272–294.

Giddens, A. (1999). *Runaway world: How globalization is reshaping our lives*. London: Profile Books.

İnceoğlu, İ. (2014). The Gezi resistance and its aftermath: A radical democratic opportunity? *Soundings, 57*, 23–34.

Kaya, A. (2017). Right to public space: Social movements and active citizenship in Turkey. *Research and Policy in Turkey, 2*(1), 1–9.

KONDA (2014). *Türkiye'de Gençlerin Katılımı*. Istanbul Bilgi Üniversitesi Şebeke Gençlerin Katılımı Projesi Kitapları - No:3, Yentürk, N., & Bahceci, D. (Eds.), Istanbul: Istanbul Bilgi Üniversitesi Yayınları. Retrieved January 7, 2022, from https://www.stgm.org.tr/sites/default/files/2020-09/sebeke-turkiyede-genclerin-katilimi.pdf.

Konrad Adenauer Stiftung. (1999). *Türk Gençliği 98: Suskun Kitle Büyüteç Altında [Turkish Youth 98: Silent Majority Highlighted]*. Retrieved January 7, 2022, from https://www.kas.de/tr/web/tuerkei/einzeltitel/-/content/turkish-youth-98.

Kovacheva, S. (2005). Will youth rejuvenate the patterns of political participation? In J. Forbrig (Ed.), *Revisiting youth political participation* (pp. 19–28). Strasbourg: Council of Europe Publishing.

Lüküslü, D. (2009). *Türkiye'de "Gençlik Miti": 1980 Sonrası Türkiye Gençliği*. Istanbul: İletişim Yayınları.

Melucci, A. (1996). Youth, time and social movements. *Young, 4*(2), 3–14.

Mouffe, C. (2013). *Agonistics: Thinking the world politically*. London and New York: Verso.

Neyzi, L. (2001). Object or subject? The paradox of 'Youth' in Turkey. *International Journal of Middle East Studies, 33*(3), 411–432.

Norris, P. (2004). *Young people and political activism: From the politics of loyalties to the politics of choice?* Paper for the conference 'Civic engagement in the 21 st Century: Toward a Scholarly and Practical Agenda' at the University of Southern California. Retrieved January 7, 2022, from https://www.researchgate.net/publication/237832623_Young_People_Political_Activism.

O'Toole, T., Lister, M., Marsh, D., Jones, S., & McDonag, A. (2003). Tuning out or left out? Participation and non-participation among young people. *Contemporary Politics, 9*(1), 45–61.

Siurala, L. (2000). *Changing forms of participation*. Unpublished paper presented at New Forms of Youth Participation, Round Table, Council of Europe, Biel. Retrieved January 7, 2022, from https://rm.coe.int/168070240f.

Sofos, S. (2017). Alone in the city: Gezi as a moment of transgression. In O. Hemer & H.-A. Persson (Eds.), *In the aftermath of Gezi: From social movement to social change?* (pp. 65–85). Cham: Palgrave Macmillan.

# 8 What is the Political? The Components of the New Post-Gezi Politics

This chapter studies how young participants of the Gezi protests relate to politics and the notion of the political. The following discussion is shaped by a number of questions: How do young people describe the political? How do they reflect on the discourse of apolitical youth? What do they think about conventional political processes and institutions? What do they base their understandings of the political on? What is the relationship between their participation in the Gezi protests and their ways of describing the political? How did their presence and experience throughout the Gezi protests affect their notions of the political and their relationship with politics? Guided by these questions, the chapter aims to understand the complex picture of Turkish youth's relationship with politics, taking the case study findings as its point of departure. To this end, the chapter examines young people's reflections on the apolitical youth discourse as well as their critical views of conventional political structures in order to contextualize the emergence of their new relationship with the political. The focus then shifts to the way the interviewees define and discuss the political in their everyday lives, cultural practices, and values, as well as to how they engage in local-level politics. This exploration of the political in young people's lives is based on ethnographic data collected from young participants in the Gezi protests, and therefore its findings cannot be applied to all young people in Turkey. However, it is important to underline that the approach to data, and the analysis here are intended to contribute to the raising of new questions and starting of new discussions on youth and politics in Turkey whose relevance would go beyond this participant group.

This chapter consists of four parts. Young people's views on the apolitical youth discourse are discussed first. This includes a discussion of their views on what it means to be apolitical along with their critique of the way this term is used against young people. Further emphasis is put on their reflections on the role that the term has played in the debates surrounding the Gezi protests. The chapter

 
P. Gümüş Mantu, *Redefining the Political: Youth Experiences of Collective Action in Turkey*, Kulturelle Figurationen: Artefakte, Praktiken, Fiktionen,
https://doi.org/10.1007/978-3-658-40565-6_8

continues with examining the interviewees' attitudes toward the traditional political arena. Their critical perceptions of politicians and the way that traditional political organizations function are scrutinized as they are seen as the root cause of their lack of trust and interest in conventional politics. This is followed by in-depth exploration of the constituents of young people's politics, identified by the study participants as everyday life relations, individual choices and local socio-cultural practices. They describe this kind of political action as one characterized by social awareness and sensitivity as well as values and humor. The last section concentrates on the organization of local forums/initiatives as a continuation of the protests as spaces where young people experimented, practiced, and continued to discover the political in the context of the Gezi protests.

## 8.1 Criticizing, Redefining, and Contextualizing the Apolitical Youth Discourse

The apolitical youth discourse in Turkey that had featured prominently in descriptions of the generation that grew up in the post-1980 period re-emerged in public debates in the wake of the Gezi protests, in which intellectuals and public figures mostly described the protests as the return of the apolitical youth to the political scene. The way young people themselves see that discourse is worth analyzing for various reasons. First, their criticism of it captures how they cope with judgmental comments from adults about their supposed apoliticism. Moreover, this discussion reveals the socio-political context these young people grew up in that limits their relationship with politics in many ways. In addition, despite their critical attitude, "apolitical" still seems to have a meaning as a concept for young people to measure their own and their peers' interest in politics. Bahar Tanyaş stresses the power of the apolitical youth discourse, referring to the degree to which it has been internalized by its subjects, that is, the young people. Although most of them do not consider themselves apolitical, they do not ignore or disagree with this discourse entirely; for example, in many cases, they point to their non-political peers (Tanyas 2015, p. 43).

It is important to highlight here that, in the following discussion, the apolitical is not treated as the opposite of the political; more precisely, political-apolitical is not constructed as a dichotomy. Rather, exploring aspects like disinterest, disengagement, or apathy is understood as a way to find the political in young people's lives. Ulrich Beck argues that the seemingly apolitical behavior of young people is actually an "antipolitics" practiced in everyday life as an area of "subpolitics":

> An ethics of everyday life is developing its own subpolitics, which is often very local and concrete and which politicians don't recognize because they don't know the cultural nerve systems of these individualized cultures. It is an 'antipolitics'. We are witnessing today an actively unpolitical younger generation which has taken the life out of the political institutions and is turning them into zombie categories. (Beck 2002, p. 230)

This section centers around young people's perception of the debates surrounding the notion of the apolitical youth and looks at young people's definitions of being apolitical, their views on where the generational and historical context of this term is derived from, and how they problematize the ways in which the notion of apolitical youth is constructed.

### 8.1.1 What Does it Mean to be Apolitical?

When discussing the concept of the apolitical youth, many interviewees (who do not consider themselves to be part of this group) described it by referring to negative attributes such as excessive consumption, individualistic behavior, and lack of interest in social problems.

> You are indifferent to social issues. What is important to you is just your own life, the ones close to you, and your freedom of consumption and travel. You remain isolated in that kind of life and you think that you are so free thanks to these opportunities. You are not interested in any kind of oppression, for example the pressures on an ethnic group in the country. (24, student, environmental activist)

Talking about the apolitical with the interviewees served as an alternative way of discovering the meaning that the political has for them. The quotation above is a good example of that. The interviewee, an environmental activist who tries to practice what he preaches by changing his consumption practices and through his way of relating to social issues and the people around him, refers to all these domains of life while talking about the apolitical.

In addition to consumption and individuality, urban life and technology are also mentioned in connection with the apolitical youth stereotype. In the quotation below, the interviewee refers to the apolitical youth as young people who mostly live in the big cities in the western part of Turkey, framing youth apoliticism as an urban phenomenon. Moreover, the emphasis on western Turkey suggests that the interviewee, who lived in the southeastern part of the country until university, excludes himself from the group:

> Apolitical youth are common in Istanbul, Ankara, or Izmir[1] because these places are where consumerism is huge, and young people have access to technology. (22, student, environmental activist)

Being apolitical is also associated with inability to question what is happening in one's environment and with general lack of critical thinking. This definition refers to young people who seem political (in that they are affiliated with political organizations or parties), but who take an uncritical, fan-like approach to politics. A leftist activist comments:

> Most of the time, being organized does not mean being political… Many young people do not question anything. In this sense, I think that many of them are apolitical. They do politics as if they were supporting a sports team; this is what being political means in Turkey. (25, employed, activist in a leftist-oppositional organization)

This interviewee also views her peers who follow their family's political traditions without question as apolitical. By this she means young people who become members of political parties or organizations because of their families' preferences or via their connections. Thinking about this tendency is also important in terms of how young people relate to their families and adults in general when it comes to the political arena. The interviewee describes these young people as being trapped in a mold:

> I know young people who are engaged with political parties, but they are apolitical, they do not think, it's family inheritance. They accept whatever they tell them and they're enclosed within a mold. (25, employed, activist in a leftist-oppositional organization)

### 8.1.2 Apolitical Youth in Their Socio-Political Context

Although many of the interviewees were critical about the notion of the apolitical youth and did not identify with the term, the discussion on the apolitical made them reflect on the context in which they grew up, and thus on how their relationship with the political developed in a different manner, in a more limited way compared to previous generations. They agree that being apolitical has its roots in the social conditions one grew up in. They admit that they are, in a sense, apolitical, and they explain their specific apoliticism as follows:

[1] The three largest cities in Turkey.

> In the '80s, people raised their children telling them: "Don't read, there's no need for this, just look out for yourself!" So, it was all about looking out for ourselves, that's the type of apoliticism we have. In fact, this is not about being apolitical, but about being alienated. To be apolitical usually means to be distant from politics, like social-asocial, political-apolitical, but no, we realize that politics exists over here, very close to us; however, we are desperate because we know that we cannot do anything. We said "to hell with it" and we focused on looking out for ourselves. (28, employed, no political affiliation)

Having been exposed to state oppression before and during the military coup, parents of the '90s generation were constantly warning their children to keep away from politics. Since their own efforts to save their country were not rewarded but severely punished, they wanted their children to "look out for themselves," which usually meant getting a good education, a well-paid job, and starting a happy family. The apolitical youth discourse was constructed against the backdrop of this type of intergenerational memory transfer.[2] The rise of these values and expectations from the individual are also the result of global processes and changing attitudes in society.[3]

The interviewees also referred to university entrance exams and the exam-based education system in Turkey as one of the factors hindering young people's political interest and intellectual curiosity.[4] The quotation below illustrates how surrounded they are by structural and social hindrances and how overwhelmed they feel both by the expectations of others and their own responsibilities. Being apolitical is, in a sense, the natural consequence of having grown up in this context:

> There is estrangement; it is not about being apolitical, but being uninterested. This is not about young people themselves... You have to do many things. You need to prepare for an exam (university entrance exam), there are many people around you talking about your life, and besides, you have your own concerns, you have to earn your own money or you'll soon have to; these things are not easy. There's always a feeling of competition... Doing well in the entrance exam was more important for us

---

[2] For more on how young people's relationship with the political, especially with organized political action, changed in the post-1980 period in Turkey, see Sect. 3.2.1. Youth Collective Action in Turkey.

[3] For more on how changing global conditions have affected young people's lives see Sect. 5.1.3. Being Young in Late Modernity: Some Reflections.

[4] The university entrance exam has been discussed in detail in connection with young people's experiences of hardships and obstacles, particularly in transition from education to work, see Sect. 6.2.2. From Education to (Un)employment: Anxiety about the Future.

> than having intellectual or political discussions with our friends. That was our apoliticism. It wasn't that we didn't appreciate these discussions, but that was the reality after the '80s. (23, student, feminist activist)

According to this interviewee, being apolitical is not about a lack of awareness; rather, young people see being apolitical as a consequence of the political, social, and economic reality they have lived through. This notion that reality is full of obstacles and difficulties goes hand in hand with their lack of hope and belief in the possibility of changing the system. This sense of despair is also pointed out as one of the reasons for the emergence of the Gezi protests:

> I was one of those young people, aware of everything, reading and things like that. But just watching... I am angry with my parents, why wasn't I born abroad? I know nothing will change. I've given up hope. I don't know whether they are apolitical, but the young are anxious and without hope... History is about repetition. Nothing is going to be better, there is just an explosion and its name is Gezi Park, it is like a crater. Young people are highly realistic. (31, unemployed, environmental activist)

### 8.1.3 Apolitical as a Concept Created by Adults

This sense of inability to change society is intensified by the older generations' pessimism about young people's ability to make a difference in the political domain. This adult pessimism also creates a hierarchy of political potentials between the generations, which ranks the older generations' political experience higher than the younger generations'. Unsurprisingly, this attitude irritates young people:

> People who lived through the '70s usually make a distinction between them and us (the youth)… They always say that they failed, therefore we won't be able to do anything either, nothing will happen, this country cannot change… My father is from this generation and he too says the same: "We did this and that, we were in jail and nothing changed, so don't push yourself too much." This makes me so angry because our youth is not the same as theirs. (23, student, LGBT activist)

They also criticize older generations' perceptions of what politics is, and their judgmental view of young people as apolitical:

> People pass judgments, such as resistance should be like this or you can oppose something in that way. I definitely do not accept that young people are apolitical. This is a very top-down attitude. We are not young in the same period as they were, we did

> not experience the same things. Therefore, we don't respond in the same way. (23, student, feminist activist)

Young people are critical of the way the term "apolitical" is defined. If it is described in reference to the conventional, adult-constructed, limited definitions of the political, it is irrelevant from the very beginning. This interviewee stresses the difference between his understanding of the political and the traditional understanding according to which he can be identified as apolitical:

> They base their arguments on some ideas according to which we are apolitical. That makes me apolitical, too. I'm not interested in your politics because it's all bullshit. I am not protesting because of this government's policies, against this government or the other one, this man in a suit or the other one. Even if the parties could change, McDonald's will not leave the country, cars will not disappear. That is why this struggle will last forever... What they can't understand is, they think we're saving the homeland, but I'm doing nothing of the kind here. (23, freelancer, independent activist)

According to this interviewee, rather than a lack of political consciousness, the apolitical means an alternative understanding of politics based on everyday life decisions, consumption practices, and values.

### 8.1.4 Apolitical Youth in the Gezi Protests

Whether it was the politically organized youth or the apolitical masses that made Gezi possible has been contested since the first days of the protests. It has been established in previous research that pre-existing groups and movements made a substantial contribution to the development of the protests and the preservation of the continuity of young people's political activism (Gümüş and Yılmaz 2015). It is commonly argued that apolitical people transformed into political actors during the protests. Reflecting on this argument, most of the interviewees who are politically organized find this interpretation simplistic and superficial. They criticize the practice of drawing strict lines between the categories of the political and the apolitical, while also questioning their own perceptions of the latter. Going beyond taking the apolitical as an essential characteristic of some young individuals, they try to change their viewpoints in order to see what is behind the discourse of youth apoliticism. One of the interviewees asserts:

> Gezi Park was not about the apolitical transforming into the political. Rather, they showed us that what we understood by politics, along with the symbols, rituals, ways of working, and some of the concrete implications we associated with it, was so inadequate and outdated and unable to reach the crowds we wanted to communicate with. The point of view that sees this process as "they were apolitical yesterday and became political today" is so metaphysical and deluded. (25, student, member of an oppositional-nationalist organization)

This statement points to the self-critical attitude of the oppositional political groups regarding their old -before Gezi- ways of practicing politics, which then perceived as hindering more people to be recruited to movements. One member of a feminist organization makes a similar comment about the possible reasons why young people who seem to be apolitical stay away from political struggles, especially street protests:

> Many women who weren't feminists visited the Feminist Tent (in Gezi Park). Their grievances were nearly the same as ours, things like the "three children issue"[5] or the abortion law, or confining young girls to their family houses. So, I think that the people who were there and are labeled as apolitical have many things to object to, but they just don't like the existing organizations in their present form. It was impossible not to see the way people around us looked at us during the protests (prior to the Gezi protests) on the street. It was like, you're doing a protest with twenty people and no one on the street is interested—this itself shows the lack of faith in that kind of protest. (28, employed, feminist activist)

One of the study participants describes the apolitical by referring to young people's tenuous relationship with the deep-rooted political and social tensions and conflicts in Turkey, such as Islamists vs. secularists, or Turks vs. Kurds, stressing the potential inherent in the ambiguous nature of the apolitical: young people may be able to be more open and tolerant toward other groups, beliefs, and ideologies, a possibility that became apparent in the relations between different groups in Gezi Park.

> Young people who are referred to as apolitical do not have a debate with the cultural codes of this society the way a Kemalist nationalist leftist in his forties or fifties does. That is to say, women with or without headscarves, people who are pious or not, or be they Kurds or Turks, they do not have strong opinions on these issues… For example, the gays and the oppositional Muslims in Gezi Park probably didn't have a marginalizing language for each other, even before the Gezi protests. (34, employed, member of an oppositional-religious organization)

[5] Referring to a speech by the prime minister, in which he said that he wanted every family to have at least three children.

While describing being apolitical, young people underline phenomena such as dependency on consumption and technology, self-centeredness, and lack of interest in social issues and critical thinking. Moreover, they see apolitical behavior as the problem of Turkey's urban west. Many of the respondents do not see themselves as apolitical. However, when they put the notion of the apolitical in the political and social context the '90s generation grew up in, they emphasize structural factors preventing them from developing an interest in politics, as well as parental pressures to stay away from it, rooted in their parents' own experiences of political oppression and violence. In that sense, being apolitical is defined as an estrangement and a conscious choice resulting from the country's peculiar social reality. The inconsistency between the older and younger generation's notion of the political is also cited as something that invalidates older generations' dismissal of the young as apolitical. The interviewees who participated in organized political groups made a visible attempt after the Gezi protests to overcome their own prejudice in order to understand those who opted out of party politics or political groups, whom they used to refer to as apolitical. They stress that being apolitical means, among other things, being free from strict traditional cultural and political positions, which indicates fresh potential for dialog between different social groups, dialog that actually happened during the Gezi protests.

In conclusion, it can be argued that, for the most part, young people criticize the apolitical-political dichotomy. They are aware that the apolitical youth discourse has a disempowering effect on young people because it dismisses their political consciousness and invalidates their unique ways of doing politics. To a great extent, they problematize the basis of the descriptions of the political and therefore regard the apolitical as a troublesome concept.

## 8.2 Politics as a Distant Realm: Too Far from Reality

When the interviewees talk about politics and the political, they usually begin by explaining how they feel about the conventional political sphere. In a way, their critical perceptions and distant attitude toward traditional party politics and representative democratic institutions, points to the ground upon which they build their alternative understanding of the political. Therefore, studying their perceptions and attitudes towards conventional politics should not be overlooked since it offers a way to understand how they imagine proper politics and the political, as opposed to the politics from which they feel disconnected.

This has been studied in detail by Demet Lüküslü in several of her works. On the basis of qualitative fieldwork conducted in 2008 with young people from various civil society organizations, she concludes that young people's critical views on politics focus on three main points: First, they see the political sphere as dirty; second, they do not believe that any change can happen in politics, and therefore feel hopeless; third, they think that political organizations lack respect for individuality and difference. Lüküslü emphasizes that these findings validate her previous field research data on young people's perceptions of traditional politics collected from 2000 to 2004 (Lüküslü 2010, pp. 193 ff.). This work follows her approach by studying young people's perceptions and critique of conventional political processes in detail, in order to better understand, rather than judge, what type of politics they imagine. My findings on how young people relate to the conventional political sphere are consistent with previous studies. How the interviewees detailed their points of critique is discussed on the following pages. Moreover, since the data discussed below was collected in the context of the Gezi protests, it offers insight into the way young people's critique of politics developed and changed throughout their protest experience.

### 8.2.1 Politicians: Lack of Trust

Most of the interviewees mention a lack of trust in the political arena as they describe it, and see politicians, political parties, and even the parliament as unreliable. An environmental activist underlines how young people's apolitical attitude reflects their lack of trust in politicians:

> I do not trust any politician a hundred percent; for me there's always at least a one percent chance that they're lying. We're discussing whether young people are apolitical, nothing that was promised to us has been delivered... Therefore, politicians can lie, and what is worse, there are people who see this as legitimate. (30, unemployed, environmental activist)

What is striking about this account is that the interviewee describes politicians' untrustworthiness as something intrinsic to them, part of their nature. As their dishonesty is taken for granted and sometimes even legitimized in society, young people do not see any room for change. Another activist states that they have no hope for it in the political sphere, and pictures politics in very strong, negative terms:

> What's happening in politics is, whoever steps into this domain ends up wading in shit. Somehow you just start doing as you're told, by the big brother or someone else. That's why today's youth have no hope when it comes to politics and they're apolitical. Many of my friends are like that… Political games are everywhere and I see politics as disgusting. (24, student, environmental activist)

Informed by their perception of the political sphere as unreliable, corrupt, and unchangeable, young people, unsurprisingly, distance themselves from politics.

> When you start thinking, reading, and talking, you also start to become politicized. But I still don't go to political protests, because I don't support such a thing... Our aim was clear there (in Gezi Park); we are against the destruction of the environment. I was there; I shouted and resisted because our goal was clear. Humankind should not destroy nature to satisfy its greed... But I don't do politics, I don't go that far, I have an idea but I will never take it to the level of politics. For example, I can't even support any political parties; I don't like what they're doing. (25, employed, no political affiliation)

Young people without political or group affiliation usually make an especially clear distinction between politics and thinking or acting on some basic issues. As the interviewee above states, she took part in the protest to express her opinion on environmental destruction, but she is careful not to designate her ideas or actions as political. Young people's hesitation to be seen as political, even when it comes to going to a protest, is very telling of the distance they put between themselves and the political sphere.

### 8.2.2 Political Organizations and Ideologies: No Room for Freedom and Critical Thought

Interviewees harshly criticize the way political organizations function and identify it as one of the main reasons for opting out of conventional political processes. They mostly see the structure of political organizations as rigid, hierarchical, unchangeable, and adult-centered. All these traditional political organizations, whether mainstream or leftist/revolutionary, are thought to limit young people's individuality, critical thinking, and freedom. One of the interviewees describes the atmosphere in political organizations as oppressive:

> I have a thing against organizations; I think they are by necessity very hierarchical and have oppressive structures that create pressure inside the organization. That's why I am pleased to stay in the gray area. I think one way on this issue, but I can change

> my mind on that other issue. I like to be in this space of freedom. (25, unemployed, no political affiliation)

As this clearly indicates, the "gray area," in which one stays at an equal distance from all ideas and decides what to do on a case-by-case basis is highly valuable to many young people. The references to the "space of freedom" or "gray area" reveal that young people opt for an understanding of political practice that strengthens rather than limits individual freedom and expression. One interviewee, who volunteered throughout her university years, explains her perception of political parties and organizations by underlining the prejudice and polarization they create. In that sense, participating in these organizations means, in a way, to "choose sides":

> I've never been a member of a political party and I've never run for any position, because what we call politics is about arranging things into patterns, and when you enter these patterns you create the opposite side. This leads to violent attitudes. It may not come to physical violence, but it creates prejudice. For that reason, I've never wanted to be part of such a thing. I know many people who think like me. However, for example, human rights, women, and children's issues are very important to me. This is indeed a political stance. (23, part-time employed, youth work volunteer)

Instead of membership in a rigid political organization, issue-based politics, focusing on specific rights, groups, and campaigns, emerges as an alternative that enables young people to be both political, in the sense of doing something about and for society, and protect their individual spaces of freedom. Ideologies, as the ground on which traditional political organizations base themselves, are perceived by one of the interviewees as instrumentalization of information. However, new -isms, such as ecological anarchism, are thought to enable free critical thinking:

> When we talk with friends who adhere to certain ideologies, they use the information presented to them to support their own ideologies. However, when I talk with eco-anarchist friends, I can feel that they question the information itself. I feel that they're not blinded by ideology, this makes me relax. I am only twenty-two years old, and to be honest, I don't want to attach myself to anything. (22, student, environmental activist)

Young people's emphasis on the importance of freedom—in terms of thinking, expressing themselves, changing their minds, and having an interest in different issues at different times—seems to have a strong link with their perceptions of being young. As they imagine youth as a period in which one has the opportunity to try out many options in many areas, including work, family and social relationships, lifestyles, and political stances, the interviewees do not want to give up their right to a test period by limiting themselves to a single political organization or ideology. Young people see membership in a political organization as a strong commitment to a specific political culture. An active member of conventional political organizations is usually thought to plan his or her life according to his or her political identity, and to prioritize politics over many other things. In that sense, being political in traditional organizations is a heavy responsibility and therefore "difficult":

> I do not see myself as politicized, because in order to be politicized I should have an organizational culture, I should be in an organization, but to be in an organization or have a political consciousness is not easy, because it's about patterns, and you have to fit in those patterns. You know what they say: He devoted his life to this and that … I think that being political is difficult. (21, student, LGBT activist)

Many young protest participants are critical about the grand discourses of traditional political organizations. One interviewee tells how she cannot identify with the discourse of "saving the country" that many of the traditional political organizations use in one way or another:

> In these kinds of [leftist] organizations, I think there's too much talk about saving the country, the homeland; it seems that there's some myth and belief... I don't know if I feel this, if I can find this love of the country in me. You see this in many groups, in different forms; nationalists love their country as well as the leftists. (24, student, no political affiliation)

This statement expresses a critical stance toward the abstract meaning of the political, that is, the myths and beliefs that do not seem to be able to connect young people to the political sphere. This sort of demand for political, national, or community belonging does not correspond to young people's understandings of life and politics. They seem to prefer a more concrete, more basic understanding of the political to these overarching concepts, and the farther a political act seems removed from practical matters and real life, the more legitimacy it loses in the eyes of many young people.

### 8.2.3 Politics: No Hope for Change

Most interviewees share the feeling that conventional politics never changes anything. Therefore, most of them do not believe that the current political system can respond to their needs and demands, or that it can make their lives better. When asked about politics, they mostly mention men quarreling on TV. On the basis of their observations and experience, they cannot imagine being able to change something by participating in conventional political processes. Moreover, many do not believe that their problems, demands, and ideas even have a place in the realm of conventional politics. One of the respondents states that she is not interested in the activities of the Parliament:

> For me, politics is something that is done on the streets at the moment. I am apolitical when it comes to parliamentary politics. What happens in Parliament is not my concern because my demands have never been considered, and it doesn't seem like they will be in the future, either. (24, student, environmental activist)

The arena of formal politics is stifling, taxing, lacks credibility and is not worth thinking about, let alone participating in, and many young people choose to spend their time and energy on things, however small they may be, that they are sure will have a political effect. An environmental activist, who spends most of her free time volunteering in a farming collective, contrasts voting with planting seeds as methods to bring about change:

> Voting is said to be the measure of one' citizenship, but how can that be possible? Sometimes it seems to me that the two seedlings that I've planted here are much more blessed. It seems voting doesn't have that much effect. That became clear after the tricks they pulled in the last elections. (24, student, environmental activist)

What is noteworthy in most of the interviews is that young people's distance from the traditional political arena is a conscious choice. Since they do not trust politicians or the mechanisms of conventional politics anymore, and do not think that the situation can change in the near future, they choose not to be a part of the system. Their attitude toward politics is not the result of ignorance, laziness, or apathy, it is, on the contrary, the product of a reflexive and critical stance. Rather than trying to enter the conventional political sphere, they define their politics through alternative areas, such as ecological activism. Most of them also clearly differentiate their experience in the Gezi protests from conventional politics, and consider their participation in the protests not political, but a humanitarian and ecological reflex.

## 8.3 Ways of Describing the Political: Everyday Life, Cultural Practices and Values

After taking a look at the interviewees' critical views of politics and the political arena, this section attempts to discover the "real" meaning of the political for young people. The following discussion examines the ideas, perceptions, actions, and practices involved in their definitions of the political in detail. Most of the interviewees' notion of the political has to do with the possibility of transformation, social change, and oppositional potential—these young people seek emancipatory politics. For many of them, the Gezi protests were a momentous experience that formed, deepened, shaped or reshaped their notion of the political. In that sense, young people's relationship with politics and their ways of describing the political deserve consideration in current theoretical discussions on the political, especially in light of the recent wave of protests and revolts in many parts of the world.

The participants in this study mostly share the conception of the political as a general consciousness about the social and political events and issues around them, and as a conscious effort to influence these events or issues. Furthermore, many interviewees describe the concept of the political as an overarching one whose manifestations can be found in many spheres of everyday life—consumption, personal relations, leisure time preferences, etc. Therefore, the political is far from being limited to a single space or institution, the political is everywhere. It is noteworthy that the foundation of this concept is described by referring to values, such as living ethically, and being conscientious and humane. The political is also described as expressing oneself, specifically via humor. Finally, all these accounts stress that the political is about action; it is practice-based, and thus cannot be understood solely through theory or discussion.

### 8.3.1 Awareness and Social Sensitivity: What is Happening to Whom in Society?

For most of the interviewees, being political refers to a basic level of awareness of one's socio-political environment:

> One doesn't need to be organized in order to be political. For example, as Nazım Hikmet said, if someone has an idea about her country and the world, understands what's happening, and follows current affairs, not only what the political leaders say but workers' deaths, occupational struggles, etc., thinks about them, presents ideas and

> talks about them from time to time, then she is a political person to me. (23, student, member of a nationalist youth organization)

While identifying awareness of what is happening around one as the very foundation of being political, this interviewee also accentuates that one should be able to analyze, reflect on, and share ideas about these processes. In this way, she frames being political as an active, conscious, and continuous type of engagement with problems and issues in society. She also stresses that being political is not the same thing as being organized formally, and therefore distances herself from the organized-political vs. unorganized-apolitical dichotomy and presents a broad understanding of the political as a type of social behavior, a point of view, a way of positioning oneself in society. Another respondent has a similar view; however, she is more focused on issue-based political thinking and action:

> For me, being political is about seeing an issue—it can be just a women's issue or a Kurdish one, minorities, any of the issues in Turkey—developing an approach and idea for that issue, trying to organize people around your idea, reflecting on it, that kind of thing. (25, employed, member of leftist organization)

On the basis of the two quotations above, it can be argued that young people who are active in more conventional organizations, such as nationalist or leftist groups, have a quite structured and concrete idea about how to be political in an organizational setting. Both interviewees underline the stages of becoming aware of a social problem, analyzing it, developing ideas, and finally, organizing struggle.

One participant, a youth worker and volunteer, sees being political as related to sensitivity to others, especially those who have different values and problems, and doing something for them:

> I can talk [about the political] on the basis of active citizenship. It is to protect, to do something about the environment we live in, about the people we live with and their values. I can define it based on sensitivity about these issues, embracing the existence of others and their values, and fighting for them. (23, part-time employee, youth work volunteer)

She emphasizes reciprocal relationships and collective spirit in society, and her way of describing the political seems to be connected with her experience as a volunteer, that is to say, being political is in a sense volunteering in the struggle against our collective problems. In addition to her previous volunteering experience, she reports that her understanding of the political deepened during the Gezi protests. The most important characteristics of relations between young people during the Gezi protests were sensitivity to others' needs and ideas, mutual protection and acts of kindness, a sense of collective belonging, and sharing, in spite of prejudices stemming from cultural and social differences. They formed the basis for collective action, thinking and discussing politics, and ultimately a redefinition of politics.[6]

### 8.3.2 Everyday Life and the Individual: Relations, Consumption, and the Local

The interviewees define the scope of the political as encompassing all aspects of everyday life. Basic daily practices such as eating, interacting with other people, nature, and animals, consumption, holiday or leisure activities, and neighborhood life are considered to be bound up with political thought and action. The interviewed environmental activists emphasize the way one constructs his or her relationship with the natural environment as an important part of the political. An interviewee explains how "natural" and "small" things became political. Power and oppression function through small natural things, such as rivers and seeds, turning them into major political issues. What is more, it turns out that people become political as a direct consequence of living in a context where their immediate surroundings are under threat. Therefore, the political is defined as a necessity, a response to present conditions. The interviewee is a member of an alternative farming group and spends a considerable amount of time every week in the garden.

> If you want to save the trees as we did in the Gezi protests, or if you want to save your rivers, you are opposing the system, because [environmental destruction] is imposed on you by the system, and when you oppose, as a matter of course you become political. The '90s generation is always labeled as apolitical, but we became political out of necessity. For example, I never thought I was participating in food politics, I realized only afterwards how valuable a seed is. Okay, it's just a seed—usually, it's like flowers or birds; you'd wonder why all these things need to be political—but in our day all these things have become political. (24, student, environmental activist)

The interviewees' emphasis on the inescapability of the political and its deep embeddedness in the most private areas of everyday life calls to mind Ulrich

---

[6] The process of collective identity formation in the Gezi protests is discussed in Sect. 4.3 Camping out in the Park: Rituals of Solidarity, Emotions, and the Gezi Spirit.

Beck's discussion of life-and-death politics. Referring to air, water, and food pollution, he points out that everyone tries to avoid harmful substances through their diet and way of life. Beck argues:

> *the most general and the most intimate things are directly and inescapably interconnected in the depths ofprivate life* in ecological culture. Private life becomes in essence the plaything of scientific results and theories, or of public controversies and conflicts. The questions of a remote world of chemical formulae intrude with deadly seriousness into inmost recesses of personal life conduct as questions of self, identity and existence and cannot be ignored. Global society is thus contained in the microcosm of private life, politics nestles down in the middle of private life and breads there. (Beck 1997, p. 152 emphasis in original).

The interviewee stresses that she was initially not aware that food could be a political topic, since it was such a normal, mundane part of life. However, her motivation to eat healthy and her wish to share this with her friends has turned into politics. In that sense, this is a mode of "life-and-death politics." Her use of the expression "our day" is interesting. Pointing to the shifting spaces of systemic threat and oppression, she also refers to the change in the meaning of the political resulting from wider socio-political change. Her account invites careful consideration of the limitations of categorizing young people as apolitical or political in trying to understand their politics.

Emphasizing basic, everyday activities in defining the political brings with it a focus on the individual. A feminist and environmental activist, who takes part in right to the city movement, insists on the individual's relationship with his or her environment:

> For me, it starts with relationships. It is about how we relate and how we live… How do we share the place we live in, what we eat, how do we form trust-based relationships? If we cannot trust each other, we cannot bring about changes in other things…We can start a political transformation by practicing [new politics] in our lives… We are severely alienated. Capitalism determines relationships, too—if you don't have any money, you can't go out with your friends. Finding out the things that can be done free of charge is therefore very political for me. (33, employed, feminist and activist in right to the city movement)

This interviewee defines the political as a bottom-up process that starts with the individual's everyday activities, positioning, and relations with society and results in the potential to bring about change. Her account shows that domains such as "the day-to-day experience, affective relations, and the deep motivations of individual behavior… become the terrain in which are bred crucial social conflicts, in which new powers and new forms of resistance and opposition confront each other" (Melucci 1996, p. 106). As the political is defined as embedded in any relation one has, be it with the social environment or food, one's political stance is elucidated as a practice deeply imprinted in the particulars of daily life. The interviewee also emphasizes how material conditions determine the content of one's life—how one spends leisure time with friends, for example. Therefore, opposition, even in the cracks of systemic oppression, such as finding activities free of charge, is an important part of being political. Related to this, consumption also emerges as one of the main spaces of political action. Another interviewee who had started to identify herself as an environmental activist after her protest experience, reports how her relationship with consumption changed:

> Life itself [is political]… When you become political in some way, it spreads to every aspect of your life. The shoes and bags I buy, my holiday preferences… For example, I don't think that I will be going on luxury holidays anymore. Consumption is everywhere, so when you become critical of consumption, you change many things. (24, student, environmental activist)

The Gezi protests saw examples of organized consumption boycotts. There were protests against many brands, shops, and restaurants, as a response to their attitude or statements regarding the protests.[7] These examples testify to the fact that consumption carries considerable potential for mass political action. One activist, who singles out the Gezi protests as a turning point in his political activity, started to identify himself as an independent activist after the protests. He started, out of political conviction, to consume leftovers from markets and restaurants, wear second-hand clothing, and travel for free whenever possible. He states that by doing so he is preventing waste, sharing his food, belongings, and all other things with other people, and challenging a consumption-oriented society. The striking point in his account is that he defines his political position as a continuous way of thinking, almost a life-long attitude and orientation. For him, the problems of today's world will not be solved by the government or revolution. Instead, an individualist, dynamic, and issue-based politics has the potential to bring about change.

---

[7] One of the biggest boycott campaigns was against Garanti Bank, a member of the Doğuş Group. Since the media branch of the group was thought to have censored news about the protests, many customers protested by canceling their credit cards and closing their cash accounts: http://www.milliyet.com.tr/garanti-1500-kart-iptal-edildi/ekonomi/detay/1718470/default.htm, accessed on May 28, 2016.

> Well, revolution will be realized by the socialists, etc. I will continue because there will still be animal exploitation, there will still be gender discrimination, buildings will be built… A movement should be focused on its issue. We were there to save Gezi Park; I wasn't trying to save a flag or anything else there. Tomorrow, I will go to defend the Northern Forests,[8] some other day, the slums. (23, freelancer, independent activist)

Individuals coming together for a cause was one of the important characteristics of the Gezi protests. Young people, especially those who were not involved in political organizations, seemed to discover the political potential of both individuals and crowds. One environmental activist, who tries to shape his everyday life in accordance with his ecological views by cycling to school and growing plants at home and in communal gardens, recounts how the Gezi protests affected people's awareness of their individual, everyday activities:

> I think people who try to do something in their individual lives are subconscious anarchists—for example, the ones who volunteer at the neighborhood garden. After Gezi, a consciousness, or rather hope emerged. People who feel that something could change and try to help the people in their neighborhoods are more hopeful and connected to each other. And this doesn't fit in boxes such as socialism, liberalism, or anarchism; it is more an individual movement, individuals becoming aware of their sense of collective belonging. (22, student, environmental activist)

This generation of young people is commonly criticized for its individualism. Against this backdrop, a discussion of the social and collective potential of the individual political act is crucially important. The Gezi protests enabled young people to see that their everyday cultural practices, their individual lives, already had political content. The political potential of everyday practices was certainly not a new concept for many young people, especially those from the feminist, ecological, or LGBT movements. For the members of these movements, the Gezi protests were an arena where they experienced and lived through these everyday practices publicly and collectively, and where their relationship with the political was deepened. For many others, the protests enabled the discovery of the political as grounded in the reality of their everyday lives.

---

[8] The Northern Forests refers to the large green areas and forests in the northern part of Istanbul. Northern Forests Defense is an ecological movement that started its activities in July 2016, just after the Gezi protests, and its main concerns have been the damaging effects of the construction of a third airport and a third bridge (named Yavuz Sultan Selim Bridge after its opening) at the Northern Forests. For more information, see the movement's website: http://www.kuzeyormanlari.org/category/english/ (accessed on September 13, 2016).

### 8.3.3 Values: Trust, Ethics, and Conscience

The interviewees often refer to values when discussing the political. In the quotations above, trust-based relationships are mentioned as an important precondition for change and being political. One of the respondents underlines that people mainly participated in the protests because of their values, such as the belief in universal human rights. In that sense, the protests were a public display of people's values. The interviewee states that:

> That kind of event occurs when something goes against people's values. And everyone there [in Gezi Park] had values and believed in human rights, universal rights. Animal rights were also an issue… People have values and they presented them there. (23, employed, youth work volunteer)

The emphasis on values in connection with the political highlights young people's distrust and critical attitude toward conventional politics which they see as lacking values such as trust, honesty, and credibility. Described in this way, the political is constructed in direct opposition to how the existing norms and practices of the formal political sphere are perceived: The political, as it is understood by young people, has a strong basis in real life and is built on trust, ethics, and conscience. In the context of the Gezi protests, the interviewees referred to conscience as the foundation of opposition to injustice. Many young people's primary motive to take to the streets was the police violence against the initial protesters that they saw on social media. It was conscience that prompted and guided them to join collective resistance, and their accounts confirm this. For example:

> The solution is your conscience. If we have people with a conscience, no matter what the problems are, we can solve them. It's a matter of conscience. Gezi was also a matter of conscience. You can solve this violence against people only by having a conscience. (22, student, environmental activist)

The emphasis on conscience is important because it helps explain the variety of political backgrounds of the protesters who came together on the basis of common values. Moreover, it would not be wrong to argue that, for young people who intentionally keep away from politics, conscience seems to function as a way to relate to the political while remaining free from ideologies and the control of political organizations. One interviewee highlights how empathy and conscience motivate young people into political action:

> We should look at the situation from a humanistic point of view. If we can offer something like this to the young, then a youth movement may appear. If it's a political ideology, it ends after a while. People move when something touches their conscience, they can empathize. They did at Gezi. (25, unemployed, member of leftist organization)

Ethics is another value-based concept referred to by young people while discussing the political. The notion of ethics undoubtedly deserves broad theoretical discussion. However, due to the focus and limitations of this study, it cannot be discussed in detail here. The study focuses on the interviewees' descriptions of it, in order to discover what they reveal about young people's relation with the political. One LGBT activist describes being ethical as the basis of the political, and it starts when one frees him or herself from systemic impositions:

> I am trying to live a life and politics means being ethical as far as possible in my relationship with life and myself. To the extent systemic impositions allow me to, I will be political. Resistance has that particular meaning for me, and this type of being political brings with it some specific actions. (27, employed, LGBT activist)

By defining it with reference to her relationship with life and herself, the interviewee demonstrates an extensive and deep understanding of the political, one that includes even an individual's inner motives and relationship to herself. Being ethical, according to her, mainly comprises the virtues of being honest with herself and others, doing what she thinks is right, in spite of the oppression; in other words, being ethical means ensuring consistency between one's beliefs, thoughts, and practices. As an LGBT activist, she confronts "systemic impositions" quite often. Resisting them turns into a struggle to be honest with herself, to live ethically, and ultimately, to be political.

### 8.3.4 "Politics on Earth": Freedom of Expression and Humor

As discussed above, young people see the conventional political arena as too distant from the realities of their everyday lives, and they do not believe that they can influence the way it functions. Therefore, politics is off limits to them, and discussing it in conventional terms is clearly pointless. However, many young participants stated that, during the Gezi protests, they started to talk about politics more and felt more at liberty to express their political views. One participant, who has no affiliation with any movement or group, explains how the protests were perceived by young people as a space where they were "able to talk":

> When these events (protests) took place, individuals one by one said, "Let's go, people are able to talk there." Politics landed on the earth for the first time; it was always up in the air before. (28, employed, no political affiliation)

Thanks to the protests, politics suddenly became relatable and accessible. Commenting on this mass politicization of the young, one interviewee points out that what some see as politicization was actually an expression of political consciousness that had formed before the protests. All the protesters needed was freedom to express themselves.

> The question isn't if people have political views or not, but if they are able to express themselves freely or not. No one could criticize the government before the Gezi protests. Everyone knows the graffiti on the walls from that period. The change happened when people started to express their views. People had something in themselves. However, they hadn't spoken out before. When they smashed the wall of fear, they started to talk. (28, employed, no political affiliation)

Walls of fear were smashed, but those of brick and concrete were young people's canvases. Humorous graffiti, along with many performative practices such as chants, music, and dance, played an important role in the protests. In her analysis of that role, Lerna Yanık argues that graffiti is in and of itself a challenge to authority, while humor multiplies that effect. Indeed, the Gezi Park graffiti aimed to deliver messages to the authorities. In addition to expressing opposition to the conversion of the park into a shopping mall, they chronicled the protesters' reaction against the officials' "paternalistic I-know-what's-best-for-you attitude" (Yanık 2015, p. 179). Young people who had not been able to interact with the conventional political sphere through customary channels shared their criticism of the government, politicians, and the policies through humorous graffiti, performances, chants, and art pieces.[9] One of the interviewees stresses the function of humor in exposing reality and establishing a continuing relationship with the political authorities and the political system, one that goes beyond the ballot box:

> In the '80s, political cartoons were much harsher. Only a few of them still exist, and they've been on trial continuously, and we've seen that the most disturbing thing for politicians is humor. If something can't be joked about, it isn't real. If you can joke about someone, that person is real. Politicians said they'd faced us with reality and this hurt us. The generation before us, my mom and dad's generation, they didn't joke about politics, they were apolitical, they said that we shouldn't get involved, they

---

[9] For a more detailed analysis of the Gezi protests from a performance studies perspective, see (Öztürkmen 2014).

> would do the right thing for us, they would vote at the polling station. But we said no! We are the follow-up mechanism. Gezi showed us this: *you* are the follow-up mechanism; *you* have to be there. (28, employed, no political affiliation)

Arzu Öztürkmen, in her analysis of the performative elements of the Gezi protests, discusses the evolution of political expression in Turkey as a background for the extensive use of humor in the Gezi protests. Given that the forms of political expression through public assembly have always been tightly controlled by the state, Öztürkmen points to other areas of expression where political debate could take place, such as the tradition of political comedy, humor magazines, or televised political debates and popular shows. She remarks that even those avenues of expression have been suppressed by the state in recent years. The creative slogans targeting the government on the roads and walls show the Turkish public's need for political humor. The suppression of political expression in its many forms seems to be counteracted through protests, young people's language of humor, play, and performances (Öztürkmen 2014, pp. 43–44 and 52). Young protesters rediscovered humor as an important way to relate to politics and define the political.

### 8.3.5 Doing Politics: The Political as Practice

Most of the interviewees' reflections on the political point to a practice-based definition of the concept. They rarely mention theorizing, long discussions, developed action plans, etc. Instead, they emphasize the here-and-now approach: doing what one can at the moment, and contributing to the transformation of society and social norms. One of the interviewees puts it succinctly:

> We don't need to talk about these things anymore. Let's make them happen; action is important. (23, part-time employed, independent activist)

He explains his practice-based notion of the political, stating the difference between discussing an idea and sharing your practical application of it. For example, being vegan, as a political stance, is not something that can be analyzed or talked about. It is a practice, and can only be understood if one engages in it, not via discussion:

> I don't want to go around and talk about being vegan to anyone else. I try to show them. When you try to explain to someone that being vegan is like this and like that,

> it's different, but when you show them what it's like, it means something else. (23, part-time employed, independent activist)

This action-based understanding of the political is undoubtedly related to young people's perceptions of the conventional political sphere. They criticize politicians for talking and doing nothing, and view the political arena as a distant realm, far removed from their reality. Therefore, they opt for an understanding of the political that they practice and apply in every aspect of their lives:

> They see politics as words, as making comments to each other. However, it's not like this. Life itself is political, it's what you wear, it's your style, your position in life, your way of communicating with a person. (20, unemployed, LGBT activist)

This politics of practice is pictured as the opposite of political ideologies. An interviewee, referring to the environment he grew up in, emphasizes that practical things, such as solidarity-based relationships or sustenance farming, can change people's lives:

> I've never defined myself as a communist, socialist, or anarchist. I've experienced too many practical things, such as growing tomatoes in pots or people sharing things with one another; I've never thought there's a need for -isms, I've always searched for something practical. My logic is that the practical is the political. (22, student, environmental activist)

Solidarity networks focusing on practical change in their areas of interest are one way young people implement this practice-based understanding of the political. One initiative collects the leftovers from supermarkets and marketplaces, cooks meals with these ingredients, and distributes them to people in need. Rejecting the obligation to pay for food on political grounds, they practice a free-food system, which underlines eating as a basic need, rather than a consumption relationship. One activist who takes part in this initiative describes this as a type of direct political action:

> All of us are in the same situation, we're either unemployed or living on too low an income. Some are also students. This is a way to live and resist wastefulness by taking money out of the equation, that is to say, this is direct action, and we're organizing it. (34, employed, right to the city activist)

Action and practice are the real side of the political and the only possible basis of change. This practice-based perspective was reinforced by the success young people achieved in preventing the construction of a shopping mall in Gezi Park.

## 8.4 An Ongoing Process of Discovery: Practicing Politics on the Local Level

Most of the interviewees describe the Gezi protests as a process through which they reflected on, discussed, or changed their notions of the political. In one way or another, young people's participation in the protests has affected the way they relate to political struggle, the political sphere, and the notion of the political itself. Following the active period of protests on the streets and the end of the Gezi Park commune (after it was cleared by the police), protest participants started to organize park forums across Istanbul, and even in other cities around the country, modeled after the protest forums. These forums were an attempt to symbolically show that any place could be Gezi Park; that is to say, if the crowds are not allowed to gather in one park, they will simply go to another. In addition, these park forums gave rise to new neighborhood solidarity organizations, local initiatives, as well as ecology and right to the city groups. Both the park forums and the nascent initiatives became spaces where people could share their understanding of the political and put it into practice. Some park forum participants interviewed here feel that the transformative potential of the Gezi protests spilled over into these forums that functioned as experimental schools of politics and free expression. The interviewees' had different experiences at the park forums resulting from the differences in their political, social, and cultural background, as well as their current circumstances. What they discovered about their relation with the political is of crucial importance for this section.

For many young people, the forums were a practice-based arena of local politics where they could discuss local issues. An interviewee, who had had no political affiliation prior to the Gezi protests, started to attend a park forum in her neighborhood and then became an activist in the neighborhood solidarity organization. Her story offers an interesting perspective on the politicization of young people in today's world. She decided to become active when she could not find an affordable apartment in her area because the rents had skyrocketed as a result of the urban regeneration process. She describes what the political means to her:

> For me, anything that affects my life is political. For this reason, I find the neighborhood solidarity organization very important. Whatever their problem is, people should try to solve it. Trying to do something about the course of one's own life is important. Very small things may be political, politics is about everyday life. (30, student, activist in a neighborhood solidarity organization)

Since politics is described as a phenomenon built into the fabric of everyday life, the space where politics is practiced mostly refers to the local arena, the narrowest circle in which it is possible to change people's lives. Neighborhood solidarity organizations as an example of local initiatives are underscored as places where people could find solutions to their problems, or in other words, do politics. Another activist from a neighborhood solidarity organization, who had been a member of the right to the city movement for a long time, emphasized the limits of representative democracy and the people's right to express themselves and have a say in the decisions that affect them.

> Politics is about letting the people express themselves, giving them authority. If you only allow people to express themselves once every five years, this is not representative democracy. I see the issue as one of rights. People should be able to make decisions about the place they live in, their education, and the taxes they pay. My understanding of the political is not about political ideologies, but about living humanely. (23, student, right to the city activist)

This quotation reveals a discrepancy between the mechanisms of representative democracy and young people's expectations from a political structure. For young people, the local bodies in which they fight to influence decision-making processes are more meaningful than political ideologies and elections every five years.

One activist, who had been a member of a leftist oppositional group long before the Gezi protests, shares that the forums were valuable for her because they helped her discover possible ways of collaborating with other people, especially those who had not yet joined a political organization. The forums provided a space in which the effect of the hierarchies based on political and organizational experience could be minimized.

> For the first time ever, there was a level playing field for collective action. Sometimes you share your experience, but they [less experienced participants] come up with many new ideas. In that sense, I can say that the forum itself changed many things in my life. (25, employed, member of oppositional leftist organization)

Another interviewee, who became an activist in a squat after the protests, makes a comparison between the traditional leftist organizations and the experiences she had in the squat. For her, the way the political was discussed and practiced by her fellow squatters was much closer to her own understanding of politics, so she preferred their practice to any organization's.

> When you join an organization, everything is actually black or white, there is no in-between. You're either a Trotskyist or a Marxist-Leninist, and there is no middle ground, there is no diversity. That place [the squat] was really a very pluralist one, and I could realize myself there as a feminist, a Leninist, a Marxist, an anarchist, and a Maoist in every form. This was a point of attraction. Secondly, it was new to us, we were learning by trial and error and that was a joy. People were so nice, it was an environment where everyone could contribute something, and there was so much motivation. No one could impose an age hierarchy, like in other organizations. You were accepted as an individual there, there were individual-based relations, it was a different life, and because of this it was so valuable. (23, employed, right to city activist)

Forums and local initiatives were not organized only in neighborhoods. Professional, white-collar interviewees also reported that forums were organized in their workplaces as a continuation of the Gezi protests. One interviewee, with no affiliation, describes how the forums she participated in at work helped her broaden her perspective on the social and political issues in the country, such as the Kurdish and Armenian identities. She states that:

> I grew up in a very sheltered environment; however, in reality, the world is this big *(she gestures with her hands)*. Turkey is this big, to say nothing of the world. I started to see many different points of view. I started to read more, to gather information from different resources. I didn't know that two of my close colleagues were Kurdish. I'd never considered this difference important, but then I started to learn about the things these people experienced. I started to listen to the stories of people from the Kurdish areas. Same with Armenians. I started to go to discussion group meetings, I started to understand these things a little bit better. As I said in the beginning, my aim was to be rich at a young age, to grow up fast. After that period, I asked myself: what is money? (26, employed, no political affiliation)

This interviewee's account underlines the transformative effect of the forum experience for young professionals. Workplace forums have the potential to facilitate different modes of conversation between co-workers and encourage them to broach topics they never discussed before. As the interviewee's account indicates, one of the main transformative effects of this environment on the participants was that it prompted them to question their relationship with money. Similar to the workplace forums, university forums were useful in bringing people together who had been isolated from each other. One university student and feminist activist reports that the forums helped her learn about the political tendencies of other students, which in turn helped her network more efficiently:

> These forums have provided us with a method, or, for example, with the knowledge of who is and isn't oppositional at the university, who is with us, who we can join forces with if something happens now and what kind of things we can do together. If a forum were to be organized now, it would be much easier—this is one of the things the Gezi protests have given us. (23, student, feminist activist)

These meetings, organized in different settings such as park forums, neighborhood solidarity organizations, workplace and university forums, and other local initiatives, helped young people form a relationship with their local spaces. Local initiatives had different agendas according to the contexts in which they developed: neighborhood solidarity organizations led collective struggle against urban regeneration, while workplace forums provided a space for critical discussion of differences and inequalities in society. As a continuation of the protests, these local initiatives are crucial for understanding how young people practice and reshape their relationship with the political in their daily lives, guided by their extraordinary experiences from Gezi Park.

## 8.5 Conclusion

This chapter discusses how the young participants in the Gezi protests perceive the political and relate to contemporary politics. Young people are highly critical of the apolitical youth discourse, especially because of the obsolete definition of the political it is based upon. However, the apolitical as a concept serves at the same time as a reference point for young people describing their own and their peers' relationship to politics. In parallel to previous qualitative studies in the field, my research shows that young people feel distant from the conventional political sphere and do not see a place for themselves in it. Their lack of interest in traditional politics is a matter of choice; moreover, their critique of political actors and institutions lays the groundwork for their own way of doing politics and defining alternative meanings of the political. In contrast to politicians, whom they view as lacking in trustworthiness and reliability, young people base their understanding of the political on trust in relationships. The political, in the emancipatory sense, understood as a potential for change in society, starts with an individual's everyday practices. They also identify relations with other people, nature, animals, and one's local environment as important parts of one's political attitude, as well as social awareness and sensitivity to problems of others, especially disadvantaged groups. Values such as empathy, conscience, and ethics formed the core of young people's understanding of the political and their political action during the Gezi protests. Humor, as a vehicle for expressing their critical opinions, is also identified as a part of political action. For most of the young participants in the protests, politics means action, and refers to here-and-now change and practice, in contrast to long debates, theoretical discussions, or long-term revolutionary plans. Local initiatives such as park, neighborhood, workplace, or university forums were the places where they reflected on their idea of the political and used it to forge a new politics after the Gezi protests.

This chapter clarifies the links between the contextual circumstances of being young and the way young people relate to the political. The effects of the process of individualization on young people's relationship with politics are mostly articulated in the interviewees' emphasis on individual choices, everyday life practices, consumption, as well as the values embodied in their individual political stances. They see it as crucially important to express themselves freely as individuals, in whatever ways they prefer, and state that the inability to do so has often made them hesitant to participate in political organizations and organized political action. They are interested in the local manifestations of global problems such as urban regeneration, environmental destruction, food and accommodation security. Under these conditions, they opt for continuous critical thinking and practice and constantly changing agendas, rather than long-term goals.

It is important to note that the Gezi protests are not taken here as the sole source of the new, emerging politics in Turkey, some distinct components of which have been discussed above. Interpreted as a moment of possibilities, the protests point to a space and temporality in which these new shifting modes of doing politics became intensified and more visible, and provide a lens for analyzing the new generation's ways of relating to the political. For many young people the protests were a watershed event that encouraged them to reflect on their experiences and perceptions of the political, and this reflective process itself offers a window into the ways they relate to politics and the political. This research demonstrates that the Gezi protests were not a "sudden" and "magical" transformation of apolitical young people into political subjects on the streets. Instead, they brought into light a change catalyzed by local and global processes that had already been underway in young people's lives, and therefore in their relation to the social and the political.

## References

Beck, U. (1997). *The reinvention of politics: Rethinking modernity in the global social order* (Mark Ritter, Trans). Cambridge and Malden: Polity Press. (Original work published 1993)

Gümüş, P., & Yılmaz, V. (2015). Where did Gezi come from? Exploring the links between youth political activism before and during the Gezi protests. In I. David & K. Toktamış (Eds.), *Everywhere Taksim: Sowing the seeds for a new Turkey at Gezi* (pp. 185–200). Amsterdam: Amsterdam University Press.

Lüküslü, D. (2010). Gençlerin Siyaset Algıları ve Deneyimleri, Yeni Bir Siyaset ve Örgütlenme Modeli Üzerine Düşünmek. In C. Boyraz (Ed.), *Gençler Tartışıyor: Siyasete Katılım, Sorunlar ve Çözüm Önerileri* (pp. 189–215). Istanbul: TÜSES.

Melucci, A. (1996). *Challenging codes: Collective action in the information age*. Cambridge, New York, Melbourne: Cambridge University Press.

Öztürkmen, A. (2014). The park, the penguin, and the gas: Performance in progress in Gezi park. *The Drama Review, 58*(3), 39–68.

Tanyaş, B. (2015). Gençler ve Politik Katılım: Gezi Parkı Eylemleri'nde 'Apolitik' Nesil. *Eleştirel Psikoloji Bülteni, 6*, 25–50.

Yanık, L. K. (2015). Humour as resistance? A brief analysis of the Gezi park protest graffiti. In I. David & K. Toktamış (Eds.), *Everywhere Taksim: Sowing the seeds for a new Turkey at Gezi* (pp. 153–182). Amsterdam: Amsterdam University Press.

Beck, U. (2002). *Zombie Categories: An interview with Ulrich Beck.* In U. Beck & E. Beck-Gernsheim, *Individualization: Institutionalized individualism and its social and political consequences* (pp. 220–231). London: Sage Publications.

# Conclusion 9

This study examines young people's relations with the political in Turkey by focusing on youth in the Gezi protests. Taking as its point of departure the question of what the Gezi protests reveal about the relationship between being young and being political, it analyzes in detail young people's experiences of the protests, their perceptions of being young, as well as their ways of defining the political, highlighting the connections between them. In doing so, this work takes a look at the way young people's perceptions of the notion of the political and their reflections on politics were informed by their experiences of being young.

Within this framework, the study has scrutinized young people's motives for protesting as well as the way they related to the collective identity built around the protests. It has addressed the socio-economic conditions of being young as well as the socio-political context shaping the experience of youth in Turkey. Furthermore, it has explored young people's relation to the political, in particular their critical stance toward the conventional political sphere, as well as what they see as "the political as it should be," crystallized in the components of the new politics which became visible during the Gezi protests. In discussing how being young influences collective action and the perception of the political, the book underscores the generational character of the politics that came into being in the context of the Gezi protests.

This concluding chapter synthesizes the overall findings of the research, presents the contributions the study hopes to make to the body of literature on the Gezi protests, highlights the questions that emerged from the research, as well as the new perspectives developed in the book.



P. Gümüş Mantu, *Redefining the Political: Youth Experiences of Collective Action in Turkey*, Kulturelle Figurationen: Artefakte, Praktiken, Fiktionen,
https://doi.org/10.1007/978-3-658-40565-6_9

## 9.1 Summary of the Research Findings and Concluding Discussion

Studying young people's experiences from 2013 revealed the background dynamics of the Gezi protests from the viewpoint of the protesters. Young individuals reported that they had been overwhelmed by feelings of unhappiness, oppression, and hopelessness prior to the Gezi protests, as a result of the impact of authoritarian politics on everyday life, the implications of neoliberal policies for the restructuring of the urban spaces, as well as instances of repressive treatment of public opposition by the official bodies. In this context, witnessing the violent treatment of the first protesters in the park was a triggering event, a moment of "moral shock" (Jasper 2011, p. 148) that led young people to take to the streets. The interviewees' accounts show that they mostly joined the protests spontaneously, without any prior organization, strategy building, or preparatory discussion. A close examination of their descriptions of the two-weeks occupation period in the park brought to light some characteristics of the collective identity that emerged during the protests, the Gezi spirit, articulated in the practices of solidarity and sharing, feelings of togetherness and trust, as well as relationships based on respect and care for others. Moreover, this period presented young people with a unique opportunity to practice their politics. They were able to organize everyday life in the park according to their ideals, and the protests therefore functioned as a laboratory for utopias. It is clear from their accounts that the protest experience caused a transformation of their everyday lives in the period following the protests, and they state that increased interest in politics and political consciousness were among the expected outcomes of the protests. The transformative impact of the protests at the individual level was mostly manifested in acts of resistance to oppression and hierarchical relationships. The interviewees stress that the way they relate to others in society has changed significantly, and that they feel a stronger sense of community after the protests. However, their perceptions of the collective identity built during the protests and their reflections on the effects of the protest experience were not homogenous. Narratives of disappointment and scepticism about the potential for collective action that emerged in the protests, as well as criticism of certain exclusionary practices directed particularly at pious protesters provide examples of different protest experiences. This work has examined young people's participation in social movements by underlining their individual experiences and paying special attention to emotions and morals. This perspective enables a deeper understanding of the process of the protests, and makes a substantial contribution to the literature on social movements, particularly in light of the recent wave of global protests between 2009 and 2014. By illuminating the role of emotions and moral values in triggering young people's collective action, and zooming in on the transformative effects of the protest at the individual level, this work adds to the studies of the Gezi protests and opens up the space for studying the link between collective action and the social conditions of being young. This is a unique contribution to the literature at the intersection of sociology of youth and the study of social movements.

The interviewees' reflections on being young show that youth as a social relationship is shaped within contextual dynamics, yet actively negotiated by young people. They describe being young with reference to having a lot of free time, energy and flexibility, being free from responsibilities, as well as having the courage for new experiences. The analysis and contextualization of these positive reflections on being young revealed that the perception of youth experience as one characterized by ability and opportunity, in accordance with the universal ideas of being young, is to a great extent contingent on socio-economic privilege, i.e. family background. Apart from this, the interviewees' accounts portrayed being young in Turkey as an experience mostly shaped by limitations and hardships. Given that the transition from education to employment is largely characterized by uncertainty, insecurity, and inequalities, as well as a growing emphasis on individual responsibility, deep and widespread anxiety about the future is reported as one of the main determinants of young people's experiences in present-day Turkey. In addition, generational conflicts, societal pressures, and gendered norms and attitudes are highlighted as factors that make youth the intersection node of several forms of power relationships. The interviewees' comparisons of being young in Turkey on one hand and in Europe on the other provide insights into what they expect from youth and how they imagine it. They emphasize that young people in Europe live in a more peaceful environment, where they could be respectful to each other, free from social pressures, and able to make their own decisions. In addition, many shared their plans to emigrate due to the political and socio-economic circumstances, which tells something about the weight of the hardships and restrictions experienced by young people in Turkey. In many instances, the interviewees define being young in relation to being political. Their perceptions of being young as being endowed with the capacity, time, and energy for political action reflect the way the discourses that attribute an innate revolutionary potential to young people are indeed reproduced by young people themselves. Young people who had already been active in political organizations and social movements before Gezi reported that age created hierarchical relationships within political organizations, and highlighted the problem of the adults' unrealistic expectations of young people in collective action. Thus, discussing being young in relation to political action underlined once more the contradictions inherent in the category of youth as it is commonly understood: on one hand, young people represent a great potential and power for transformative political action, while on the other hand, they are often manipulated, oppressed, and treated as a group that needs to be lead and directed.

The protests were described as enabling a change in the way the participants experienced being young, and one of their immediate effects was that the older generations' judgmental and condescending perceptions of the political potential of the young generation lost all credibility after the protests, which made young people feel much more recognized and respected in society. In contrast to being young in an "ordinary" way, the interviewees' accounts presented the protests as an extraordinary context in which being young became an easier experience, free from restrictions, hardships, and oppression. The protests, especially the two weeks of camping in the park emerged as an exceptional time and space that enabled young people to express themselves freely, have their needs met through solidarity-based organization of everyday life, develop empathetic relationships with others, and practice and perform their political ideals. All of this they did with fun and joy despite the constant threat of police intervention. Thus, the Gezi protests turned out to be a "youthful" (Bayat 2013, p. 18) space for young people, where they freed themselves of the limiting and oppressive dynamics shaping the experience of being young in ordinary life, where being young became a liberating experience. This work defines the Gezi protests as a "youthful" experience, underlining the transformation of the ways of experiencing youth they made possible. Making a conscious effort to avoid reproducing essentialist stereotypes of young people, this work uses the term "youthful" to define the exceptional modes of being young in the protests by emphasizing how young people themselves describe the protest environment as one in which they felt younger and therefore youthful. Thus the investigation of the notion of youth in this book not only makes a contribution to the sociological understanding of young people's lives in Turkey, but also offers new perspectives for studying youth in the context of protest and in relation to the political.

Young people's reflections on the concept of the political operate against the background of their critical discussions of the notion of the apolitical and the functioning of conventional political sphere. The interviewees drew attention to the fact that apoliticism was actually a label used by the older generations to describe young people's political attitudes, and it reflected a condescending attitude toward young individuals that disregards all the multifarious new ways in which they related to the political. However, they do not dismiss the term completely; rather, they use it frequently to differentiate themselves from other young people who are not interested in social issues, which lends the label some explanatory potential. They also use it in an affirmative way, to refer to the young generations' positive habit of rejecting the confines of outdated political ideals and polarizations. The findings of the research of young people's perceptions of traditional politics match the results of previous examinations of this topic. Young people express distrust toward politicians, they see conventional political organizations as restrictive and hierarchical structures leaving no space for critical thinking, individuality, and freedom, and as they have no hope in the transformation of traditional politics, most of the interviewees did not see it as worthwhile or even possible to step into the traditional political arena.

The interviewees' definitions of the political stand in contrast to their perceptions of conventional politics. Contrary to the image of high politics as faceless, young participants describe the political from their viewpoints with references to everyday life, relations between individuals, values, and practices. The political is first of all associated with developing an awareness of what is happening in society as well as taking action and getting involved. Relations with others and the ways one relates to one's social and natural environment are underlined as illustrating one's political position, while adopting ethical consumption practices by paying attention to how consumption creates social stratification and organizes social life, and behaving consistently with one's values and ethics is accentuated as a mode of political action. The interviewees' accounts demonstrate that expressing oneself freely, in one's preferred way, for instance through humor and creativity, forms the backbone of politics for young people. Moreover, the political, as described by the interviewees, exists in the realm of the practical. Theorizations, discussions or strategy building are not wholly rejected, but they are meaningless if viewed separately from action. On the other hand, oppositional politics and transformative action are in a symbiotic and dialogical relationship, e.g. ecological activism is impossible without engaging in critical evaluation of one's personal practices that affect one's surroundings and taking action to achieve individual transformation. The close exploration of their relationship with the political and their ways of defining the political in this work underlines how young people's politics are shaped by the effects of structural processes such as individualization and globalization, in addition to the specific contextual dynamics and social conditions of being young. By highlighting their perspectives in describing the political, this work proves the importance of developing a youth-informed understanding in analyzing young people's relationship with the political.

What this research has demonstrated is that the social conditions as well as the social construction of being young are bound up with young people's perceptions of the political and their ways of taking part in collective action. The Gezi protests, as an extraordinary spatiality and temporality, illustrated how young people's contextual experiences and their political action were interrelated in various complex ways. The "youthful" experience of the Gezi protests gave young people an opportunity to step out of their everyday lives in which they had suffered social pressures, oppressive intergenerational relations, and financial dependence on their families. Furthermore, Gezi was an occasion for them to stand up against the repressive Islamization and neoliberal policies, and a free space outside of the control of the hegemonic non-secular and market-oriented order.

The interviewees' emphasis on the importance of individual political action revealed something about their experiences of individualization in various spheres of everyday life. Faced with structural problems ranging from education to employment, health, and housing, young people seem to adhere to a sense of individualism even when engaging in political action. Although they acknowledge the value and importance of collective identity (indeed, they created one during the protests), they find it crucially important to retain their individual focus and critical attitude. This individualism is at odds with the culture of traditional political organizations and makes it nearly impossible for young people to engage in politics via traditional channels. Their insistence on describing their participation in the Gezi protests as individual—even if some of them were members of political organizations—was one of the examples of this tendency.

Growing up in a specific historical and social context in Turkey in the 2000s, in other words, sharing a generational location, the participants of the protests were exposed to similar processes of oppression and experienced similar hardships. The protests enabled young people to become aware of their generational commonalities. While collective exposure to contextual pressures and difficulties was the background to their participation, they developed a collective consciousness as they experienced a stimulating synchronization of emotions and moral values as members of the same generation. This research has argued that this generational consciousness was the basis of young people's collective identity and action in the Gezi protests. Free from the rigidity of the political organizations, they found in their generational commonality a broader and inclusive space of collective action where individuality, emotions, values, relationships, humor, and creative performances were all at play in the creation of political and transformative experiences. Generational consciousness was thus an important constituent of the youthful space created in the protests. As they incorporated a critical stance toward the function of the category of youth in society into their political action, their new ways of engagement with the political pointed to possibilities for developing an age/youth-conscious politics. Tracing the ways in which being young is mediated thus offers new insights into the relationship between young people and politics that go beyond the labels of "revolutionary" or "apathetic" youth.

This book interprets young people's participation in the Gezi protests as the result of the ways in which being young was experienced in Turkey in the 2000s, and reveals the way in which young people's relationships with the political shaped the character of the protests. The protests were a momentous event, yet one limited by time and space, thus a definition of the political and a discussion of politics that is applicable to the Gezi protests only is neither possible nor valid. Therefore, this book treats the protests as a mechanism through which young people's nascent politics became temporarily tangible. As has been argued in several places in this work, the transformation of young individuals' relationship with the political is affected by several local and global social processes and is itself an ongoing process. It is thus important to highlight that this research was a snapshot of youth politics during the Gezi protests, and can therefore shed only limited light on young people's relation to politics in general. However, the most notable finding is the fact that the protest experience has led to a transformation in several spheres of young people's lives, especially their ways of being young and relating to the political, and this transformation is expected to have an effect on their future relationship with collective action and the political domain.

## 9.2 Limitations of the Study and Notes for Future Research

This work is based on the field research conducted with a group of young participants of the Gezi protests in Istanbul. Given that the protests eventually spread over nearly the whole country in a short span of time, this work is obviously limited since it does not look at other parts of the country. As stated in Chap. 2, it does not aim to be representative but to develop an in-depth understanding of its subject matter using the ethnographic approach. Therefore, the limitation is not about representativeness, but rather refers to the homogeneity of the study participants' experiences in terms of geographical distribution. However, it is also important to underline that narrowing the focus down to the participants of the protests in Istanbul was a deliberate decision taken in accordance with the methodology and questions of the research. Focusing on Istanbul was important for two reasons, namely collecting the accounts of young people who camped in the park during the two-week occupation period, and collecting their observations and interpretations of the events following the initial protests at Gezi.

The inquiry into the generational dimensions of the protests was centered around the interviewees' accounts that defined generational experience mostly with reference to collective exposure to specific historical and structural influences as well as the ways in which young people negotiated their way through these contextual processes. Their creative and performative practices that shaped the protests are not included in the analysis here as the in-depth ethnographic interviews were focused on young people's individual perspectives. However, it is worth underlining that these cultural practices manifested in the chants, slogans, and social media posts point to the cultural aspects of the generational consciousness that emerged in the protests, and demonstrate that young people actively engage with the social and cultural material of their times and shape their generational identity. Therefore, an analysis of their cultural and performative practices, in conjunction with an examination of their individual reflections, would undoubtedly yield deeper insights into the Gezi protests.

This work has studied young protest participants, thus it mainly reflects the politics of young people who hold oppositional views regarding the status quo in the country. While a detailed examination of the perspectives of young people who were against the Gezi protests is outside the scope of this research, it could make a valuable contribution to the analysis of the Gezi protests. In addition, continuous reconsideration of young people's relationship with politics seems to be a necessity, given the rapidly changing political dynamics in Turkey. How was young people's engagement with the political shaped in the face of increasingly authoritarian policies? How was political polarization, mostly evident in the tensions between secular-oppositional and religious-conservative sections of society, reflected in young people's perceptions of society and politics? Research focused on the influence of socio-political dynamics in Turkey on the complex ways in which young people from different backgrounds relate to the political would certainly broaden our understanding of young people's politics as well as the conditions and construction of being young.

This research was finalized in August 2018, five years after the Gezi protests, and it is now important to explore if and how young participants' views have changed over time, take a look at the long-term effects of the protest experience on young individuals and trace the evolution of their politics over the years. Are young people capable of creating youthful political spaces in which they can express themselves and resist the oppressive dynamics of being young in Turkey? Given that the opportunities for collective oppositional action are becoming increasingly limited in the current political atmosphere in the country, young people's politics are becoming less and less visible in the public space, and we see fewer and fewer instances of mass protests and spectacular forms of collective action. At this point, the question of what happened to the political possibilities and the transformative potential that emerged in the exceptional spatiality and temporality of the Gezi protests stands as one of the most critically important lines of inquiry. Looking back at the findings of this research, an in-depth investigation of young people's practices in everyday life and micro-spaces promises insights into their constantly changing but vital politics that probably persist in latent forms.

## References

Bayat, A. (2013). *Life as politics: How ordinary people change the Middle East*. Stanford: Stanford University Press.

Jasper, J. M. (2011). Emotions and social movements: Twenty years of theory and research. *The Annual Review of Sociology, 37*, 285–303.

MIX
Papier aus verantwortungsvollen Quellen
Paper from responsible sources
FSC® C105338

If you have any concerns about our products,
you can contact us on
**ProductSafety@springernature.com**

In case Publisher is established outside the EU,
the EU authorized representative is:
**Springer Nature Customer Service Center GmbH**
**Europaplatz 3, 69115 Heidelberg, Germany**

Printed by Libri Plureos GmbH
in Hamburg, Germany